Asheville-Buncombe
Technical Community College
Learning Resources Center
340 Victoria Road
Asheville, NC 28801

Volunteer Training Drills

DISCARDED

JUL 2 5 2025

Volunteer Training Drills

A Year of Weekly Drills

Howard A. Chatterton

> **Disclaimer**
>
> The recommendations, advice, descriptions, and methods in this book are presented solely for educational purposes. The author and publisher assume no liability whatsoever for any loss or damage that results from the use of any of the material in this book. Use of the material in this book is solely at the risk of the user.

Copyright © 1998 by Fire Engineering Books & Videos,
a Division of PennWell Publishing Company.

All rights reserved. No part of this work covered by the copyright hereon may be reproduced or used in any form by any means—graphic, electronic, or mechanical, including photocopying, recording, taping, or information storage and retrieval systems—without prior written permission of the publisher.

Published by Fire Engineering Books & Videos
A Division of PennWell Publishing Company
Park 80 West, Plaza 2
Saddle Brook, NJ 07663

BOOK DESIGN BY MAX DESIGN
COVER DESIGN BY STEVE HETZEL

Printed in the United States of America

Library of Congress Cataloging-in-Publication Data

Chatterton, Howard A., 1943-
    Volunteer training drills / by Howard A. Chatterton.
       p.   cm.
    ISBN 0-912212-68-3 (softcover)
    1. Fire fighters—Training of. 2. Fire drills. I. Title.
TH9120.C49 1998
628.9'25'0715—dc21   98-21540
                                                                                                      CIP

About the Author

■ Howard Chatterton joined the Bowie (MD) Volunteer Fire Department and Rescue Squad Inc. in 1968, and he is a life member of that department. He became deputy chief for rescue and a fire suppression captain. He is a state-certified emergency services instructor and a field instructor for firefighting courses at the Maryland Fire and Rescue Institute. He is coauthor of *Making and Managing Money—A Financial Planning Workbook for Volunteer Fire Departments* (Fire Engineering Books and Videos, 1994).

To my wife, Muffet, and our daughters, Christine, Jacqueline, and Kathleen. Thanks for all your love, support, and encouragement.

To the members of the Bowie Volunteer Fire Department and Rescue Squad. Thanks for your efforts to train, drill, and provide expert fire suppression, rescue, and emergency medical service to the City of Bowie, Maryland.

Table of Contents

Introduction ix
I. Planning Your Drill Program 1
II. Table of Drills 5
III. Guidelines for the Training Officer 7
    Drill Safety Officer's Checklist 9
IV. Drill Outlines 11
    1. Driver Proficiency 11
    2. Ropes and Knots 16
    3. Roof Operations 19
    4. Ground Ladders 27
    5. Rescue by Ladder 33
    6. Preplans 37
    7. Physical Fitness I 41
    8. Physical Fitness II 50
    9. SCBA I 54
    10. SCBA II 59
    11. Explosive Devices 62
    12. Arson Investigation 64
    13. CPR/AED 67
    14. Forcible Entry—Conventional 69
    15. Hose Handling I 74
    16. Pipeline Emergencies 77
    17. Electric Utility Emergencies 79
    18. Hose Testing 81
    19. Vehicle Extrication 84
    20. Foam Hoselines 88
    21. Elevated Stream Operations 91
    22. Engine Company Evolutions 94
    23. Residential Fires SOP—Combined Evolutions 97
    24. Advancing Hoselines by Ladder 105
    25. Drowning Response 108
    26. Swimming Accidents 112
    27. Area Familiarization 115
    28. Apparatus and Equipment 117
    29. Ventilation 120
    30. High-Rise Response 125
    31. Multiple-Casualty Incidents 130
    32. Haz-Mat Operations With Multiple Casualties 134
    33. Railroad Emergencies 139
    34. Military Aircraft Emergencies 141

35. Hose Handling II 143
36. Forcible Entry—Through-the-Lock Methods 146
37. Medevac Helicopter Operations 150
38. Burns 152
39. Fractures 157
40. Soft Tissue and Penetrating Wounds 159
41. Cold Weather Emergencies 161
42. Hot Weather Emergencies 166

Figures

List of Contacts 3
Drill Safety Officer's Checklist 9
Apparatus Maneuvering Course (Fig. 1-1) 15
Roof Operations Quiz (Fig. 3-1) 22
   Quiz Answers (Fig. 3-2) 23
Peaked Roof Collapse—Lecture Outline (Fig. 3-3) 24
Checklist for Ladder Inspections (Fig. 4-1) 30
Checklist for Ladder Skills (Fig. 4-2) 31
Rescue Dummy (Fig. 5-1) 36
Quick Access Preplan (Fig. 6-1) 40
Health Risk Evaluation Form (Fig. 7-1) 45
Stretching Exercises (Fig. 7-2) 46
Navy Physical Readiness Standards (Fig. 7-3) 47
Personal Fitness Goals and Progress Chart (Fig. 7-4) 48
Flexibility Board (Fig. 7-5) 49
Room Arrangement for SCBA Drill (Fig. 9-1) 57
SCBA Skills Checklist (Fig. 9-2) 58
Arson Investigation Checklist (Fig. 12-1) 66
Forcible Entry Quiz—Conventional Techniques (Fig. 14-1) 72
   Quiz Answers (Fig. 14-2) 73
Hose Handling I (Fig. 15-1) 76
Hose Testing Checklist (Fig. 18-1) 83
Vehicle Extrication Evaluation Points (Fig. 19-1) 87
Foam Operations Checklist (Fig. 20-1) 90
Drill Announcement Flyer (Fig. 23-1) 104
Advancing Hoselines Checklist (Fig. 24-1) 107
Tool Inspection Checklist (Fig. 28-1) 119
Hose Handling II (Fig. 35-1) 145
Adams Rite Lock Demonstrator (Fig. 36-1) 149
Illustrations for Burns Quiz (Fig. 38-1) 154
Burns Quiz (Fig. 38-2) 155
   Quiz Answers (Fig. 38-3) 156
Cold Weather Emergencies Quiz (Fig. 41-1) 163
   Quiz Answers (Fig. 41-2) 165
Hot Weather Emergencies Quiz (Fig. 42-1) 168
   Quiz Answers (Fig. 42-2) 169

Introduction

Chief Ramon Granados of the Bowie (MD) Volunteer Fire Department and Rescue Squad often said that "The public will probably forgive you if you have trouble with a difficult fire, but they will never forgive you if you don't do the basics right."

If your members cannot get water on a first-floor fire within a minute of arrival, or if they cannot throw a ladder to the second floor and make a rescue, the public will not forgive you—and they probably shouldn't!

Training is all about learning new skills and certifying the ability to perform them to a given standard. Some of the outlines in this book are concerned with setting up training for your members. The majority of the outlines provide guides for a series of drills that will challenge your members; provide an opportunity to measure their progress; and, where possible, incorporate the fun of competition. The goal is to practice the basic skills until they become automatic.

Section I provides a suggested one-year outline for the training officer (TO). With a little advance planning, the TO can establish a weekly one-year program with minimal effort.

Section II contains a table listing all of the outlines in this book. It gives a quick summary of location (inside or outside) and the amount of preparation associated with each drill.

Section III contains some general guidelines for the training officer. Read these guidelines. They contain the basics that you must follow if you are to do your job right. Neither your members nor your chief officers will forgive you if a member or spectator is injured because you failed to follow the basics.

Section IV contains the outlines. They have all been tested, and they work. There is nothing sacred about them, however. Customize them to your department's operations, but remember, there are no excuses for a member being injured due to a poorly thought-out drill.

I. Planning Your Drill Program

The table in Section II lists the characteristics of the drills in this book. Get out a calendar and start laying out the year's drills. Once each quarter, you should schedule a driver-training drill and a live-burn exercise at your local training facility. The following scheduling recommendations are based on those areas of the country that experience four distinct seasons.

Fall—13 Weeks

1. Driver Proficiency—Drill No. 1.
2. Advancing Hoselines by Ladder (Live-Fire Exercise)—Drill No. 24.
3. Ground Ladders—Drill No. 4.
4. Rescue by Ladder—Drill No. 5.
5. Physical Fitness I—Drill No. 7.
6. Physical Fitness II—Drill No. 8.
7. Soft Tissue and Penetrating Wounds—Drill No. 40.
8. Roof Operations—Drill No. 3.
9. Elevated Stream Operations—Drill No. 21.
10. Ventilation—Drill No. 29.
11. Multicasualty Incidents—Drill No. 31.
12. Vehicle Extrication—Drill No. 19.
13. Residential Fires SOP—Drill No. 23.

Winter—13 Weeks

1. Driver Proficiency—Drill No. 1.
2. Cold Weather Emergencies—Drill No. 41.
3. Drowning Response, Ice Rescue—Drill No. 25.
4. Fractures—Drill No. 39.
5. Ropes and Knots—Drill No. 2.
6. Explosive Devices—Drill No. 11.
7. Arson Investigation—Drill No 12.
8. CPR/AED—Drill No. 13.
9. Pipeline Emergencies—Drill No. 16.
10. Electric Utility Emergencies—Drill No. 17.
11. Burns—Drill No. 38.
12. Apparatus and Equipment—Drill No. 28.
13. Forcible Entry, Conventional—Drill No. 14.

Spring Drills—13 Weeks

1. Driver Proficiency—Drill No. 1.
2. Forcible Entry, Through-the-Lock Methods—Drill No. 36.
3. SCBA I—Drill No. 9.
4. SCBA II—Drill No. 10.
5. Physical Fitness I—Drill No. 7.
6. Hose Handling I—Drill No. 15.
7. Medevac Helicopter Operations—Drill No. 37.
8. Preplans—Drill No. 6.
9. Engine Company Evolutions—Drill No. 22.
10. Advancing Hoselines by Ladder (Live-Fire Exercise)—Drill No. 24.
11. Railroad Emergencies—Drill No. 33.
12. Residential Fires SOP—Drill No. 23.
13. Area Familiarization—Drill No. 27.

Summer Drills—13 Weeks

1. Swimming Accidents—Drill No. 26.
2. Foam Hoselines—Drill No. 20.
3. Advancing Hoselines by Ladder (Live-Fire Exercise)—Drill No. 24.
4. Elevated Stream Operations—Drill No. 21.
5. High-Rise Response—Drill No. 30.
6. Hot Weather Emergencies—Drill No. 42.
7. Haz-Mat Operations—Drill No. 32.
8. Physical Fitness II—Drill No. 8.
9. Hose Handling II—Drill No. 35.
10. Hose Testing—Drill No. 18.
11. Vehicle Extrication—Drill No. 19.
12. Military Aircraft Emergencies—Drill No. 34.
13. Engine Company Evolutions—Drill No. 22.

PLANNING YOUR DRILL PROGRAM

List of Contacts

With a basic schedule set, get on the telephone to schedule the people you want to come in and give drills. Record the names and phone numbers below to save time when you need to contact them to confirm dates and times.

Item (Drill No.)	Contact's name	Contact's title	Contact's phone no.	Date to call back and confirm
Road cone supplier (Drill No. 1)				
Bomb squad representative (Drill No. 11)				
Arson investigator (Drill No. 12)				
CPR instructor (Drill No. 13)				
Electric utility representative (Drill No. 17)				
Pipeline utility representative (Drill No. 16)				
Swimming pool manager (Drill No. 26)				
Railroad safety representative (Drill No. 33)				
Military aircraft representative (Drill No. 34)				

4 Volunteer Training Drills

Your list of contacts is a valuable asset. These people can provide brochures and handbooks for your members that will enhance your effectiveness as a training officer. Treat your contacts well. Call them a week before their drill to see whether there is anything that they need in the way of logistics or supplies. At the end of the drill, give them a company patch or coffee mug. When they get new information or supplies that might be useful to you, they will remember you.

With a schedule in hand, read through each of the drills. Some can be set up very quickly. Some require obtaining permission to use local buildings or facilities. A few require building simple props. The high-rise and mass-casualty drills require some advance planning.

Look up the references. Make copies of the articles that provide the supporting materials for the drill. Put the copies into a notebook so that it becomes a valuable training reference tool. Keep up with your professional reading. As you find new articles on the drill subjects, add them to the notebook, and update the drills to keep pace with changes in policy, technique, and equipment.

Each drill ends with space to note what went right or wrong and what should be done to improve it the next time around. Don't become a slave to your schedule. If the members have trouble with one of the drills, repeat it to get to the desired skill level.

Few members want to drill on the basics. That is why these drills are so crucial. Chances are that you will find that the basic skills are a bit rusty, especially in companies that don't have a high incidence of fires. Few members like discovering that their skills have deteriorated. Hang tough and develop a thick skin. You can never go wrong by improving the basic skills of your members.

II. Table of Drills

Drill Topic	Drill No.	Outside station	Inside station	Preparation required	Coordination required	EMS involved
Advancing hoselines by ladder	24	X		Little	Training center	Optional
Apparatus and equipment	28		X	Little		Yes
Area familiarization	27		X	Little		Yes
Arson investigation	12		X	Little	Investigators	Optional
Cold weather emergencies	41		X	Little		Yes
Combined evolutions	23	X		Moderate		Yes
CPR recertification	13		X	Little		Yes
Driver training	1	X		Moderate		Yes
Drowning	25	X	X	Little		Yes
Electric utilities	17		X	Little	Utility company	Optional
Explosive devices	11		X	Little	EOD unit	Yes
Foam hoselines	20	X		Little		
Forcible entry (conventional)	14	Optional	X	Moderate		Yes
Forcible entry (through the lock)	36		X	Moderate		Yes
Fractures	39		X	Little		Yes
Gas utilities	16		X	Little		Yes
Ground ladders	4	X		Little		
Haz-mat SOPs	32	X		Moderate		Yes
Hot weather emergencies	42	X		Little		Yes
Helicopter operations	34, 37	X	X	Little	Medevac/ military	Yes
High-rise response	30	X		Much		Yes
Hose handling	15, 22, 35	X		Little		

5

Volunteer Training Drills

Drill Topic	Drill No.	Outside station	Inside station	Preparation required	Coordination required	EMS involved
Hose and hydrant tests	18	X		Little		
Incident command	30, 31, 32	X		Moderate to much		Yes
Ladder pipe	21	X		Little		
Live burn—structure	24	X		Little	Training center	Optional
LPG tank fires	15	X		Little		
Map book	27		X	Little		Yes
Military aircraft	34	X	X	Little		
Physical fitness	7, 8	X		Moderate		Yes
Preplans	6		X	Little		Yes
Railroad emergencies	33	X	X	Little	RR safety office	Yes
Residential fires	23	X		Moderate		Yes
Rope and knots	2	X		Little		
Search and rescue	9, 10		X	Little		
Soft tissue and penetrating wounds	40		X	Little		Yes
Swimming pool accidents	26	X		Moderate		Yes
Triage	31	X		Moderate		Yes
Vehicle accident—extrication	19	X		Moderate to much		Yes
Ventilation	29	X		Moderate		

III. Guidelines for the Training Officer

As a training officer, you have the opportunity to greatly improve the efficiency of your department by developing the skills of your members. You must think through every drill evolution, follow every applicable standard, and ensure the safety of your members and the public. A drill is a controlled situation. You must provide that control.

The Basics

1. Use the incident command system. If you use the ICS routinely, it will become automatic in your operations.

2. For every drill, establish as a standard procedure that a whistle blown by a training or safety officer is a signal for everyone to stop in place.

3. Establish a policy that any incident of horseplay during a drill is cause for formal disciplinary action. Be prepared and willing to enforce this rule.

4. Outside evolutions attract spectators, especially children. Designate an area for spectators so that they will be out of the path of moving apparatus, burst hoselines, hose streams, falling ladders, and other hazards. Use barrier tape to outline a spectator area, and warn your members to watch for those who might wander into danger.

5. Think about the location. Where are you holding this drill? Are you in plain view of a major intersection or a heavily traveled road? Are you going to distract drivers? Are you liable to be the cause of an accident?

6. Think about the location again. Is your ladder pipe or master stream operation going to damage property? Where is the water going to drain? Are you going to scour those freshly painted parking lot lines right off of the pavement?

7. Think about the weather. Is the temperature going to drop? Is your evening hose evolution going to turn a shopping center parking lot into an ice rink by morning?

8. Assign a safety officer (SO). Provide the SO with the checklist in this section for monitoring the drill. Consider this checklist to be a starting point—then improve on it! Make a copy of the checklist and use it at every drill.

9. Have a plan for each drill and do not deviate from the plan. If the members want to try something new, STOP. Think it out, and either formulate a new outline on the spot or defer to the next drill so that you can create a proper plan. Do not freelance, and resist the pressure from your members to do so. Remember, if something goes wrong, you will be explaining why you deviated from a planned program to try an unplanned experiment.

10. Think about the hazards inherent in the drill. What could go wrong, and what steps will you take to minimize those hazards?

11. Read the outline and the reference material ahead of time. Be prepared and organized so that your members aren't standing around waiting for you.

12. Live burns should only be conducted in a fire training facility under the guidance of certified instructors. If you don't have access to such a facility and must conduct your suppression training using donated buildings, you must first read the National Fire Protection Association Standard 1403, *Standard on Live-Fire Training Evolutions*, and follow those recommendations to the letter. These are consensus standards, not laws. However, should an injury occur and legal action be taken, your liability as a training officer will be judged against those standards. Another good reference is "Conducting Safe Live Burns" by Don McFeely, *Fire Engineering*, June 1993, page 35.

13. Consider posting signs reading "Fire department training," both to inform the public and as a public relations tool. If you are conducting live-fire evolutions in acquired structures, pass out brochures in the immediate neighborhood informing the public of the exercise. This is a good public relations tool, and it will lessen the number of people reporting a fire or smoke in the area.

14. Have a backup drill ready to go in case of a problem with the weather, the location, or your expected guest speaker.

Drill Safety Officer's Checklist

Weather
Consider temperature and humidity.
    Provide water and cups during the drill.
    Ensure that a rehab area is available, whether the weather is hot or cold.
    Ensure that hose evolutions will not create icing in public areas.
    Ensure that there is no smog alert in effect during outside evolutions.

Safety Apparel
The members must wear and use approved gear, especially:
    Face shields, which must be in place.
    Department-issued gloves.
    Department-issued boots.

Tools
The members must handle tools correctly, especially:
    Pike poles.
    Axes.
    Power saws, which must be turned off when climbing.
    Tools in general must be carried correctly when climbing ladders.

Apparatus
Vehicles must be placed properly:
    Stable ground.
    Wheel chocks in place when parked.
    Traffic cones placed to protect the apparatus and personnel.

Spectators
Have a designated spectator area that is out of line of moving apparatus, hose-line failures, hose streams, falling ladders, and other potential hazards.

Watch the perimeters, especially for small children.

Volunteer Training Drills

Member Safety

Pay attention to members who are fighting colds or who have recently experienced illness.

Ensure that the members hydrate prior to drills in warm or hot weather and that they rehydrate between evolutions.

Ensure that the members take a break between evolutions.

Ensure that there is EMS support or that first-aid and oxygen kits are on hand.

Ensure that the lighting is adequate.

Lifting and Ladders

Ensure that the members lift with their knees bent and their backs straight.

Ensure that they carry ladders correctly.

Check for overhead obstructions and wires before raising ladders.

Ensure that the ladders are tied in.

Additional Items

Do not flow large quantities of water if local water restrictions are in effect.

Do not conduct burns if local air pollution alerts are in effect.

Do not let the members continue to do something wrong and then criticize them for it. Stop the action, demonstrate the proper technique, then continue the drill. Perfect practice makes perfect performance.

Assign additional safety officers as needed.

IV. Drill Outlines

1. Driver Proficiency

Objective

To develop skills in maneuvering vehicles in close quarters.

This is a good drill to do once every quarter. Keep a record of those drivers who participate. If you do have an accident, you can produce a log of driver training.

Setup Time

First time: Two hours.

Subsequent times: 30 minutes.

Materials Required

Authorization letter to use the property.

75 road cones (minimum 18 inches tall; 24 inches preferred).

50- or 100-foot tape measure.

Pavement marking paint.

Maps of the course.

Two portable radios on the same frequency as apparatus radios.

Reflective clothing for the course workers if training at night.

Wand-type traffic control flashlights if training at night.

References

"Air Brakes and the Driver-Operator," Terry Eckert, *Fire Engineering*, March 1998, page 21.

"Tanker Operations: Special Handling Required," Dale Perry, *Fire Engineering*, March 1998, page 157.

"New Technology for Emergency Vehicle Sirens," Barry McKinnon, *Fire Engineering*, November 1990, page 65.

"Defensive Responding," Vincent Dunn, *Fire Engineering*, August 1989, page 44.

NFPA 1002, *Fire Department Vehicle Driver/Operator Professional Qualifications*.

Preparation

1. The first step is to find a parking lot big enough to accommodate the course. The best bet is a school parking lot, preferably one with lights for night use. Be sure to contact the property owners or the local school board to obtain permission to use the property. Also, get permission to put small spots of paint on the pavement so that you can set up the course again quickly.

2. You will need about 75 road cones. If your department doesn't have a supply of them (about five dollars apiece from an industrial supplier), you can usually borrow them from your highway department or perhaps a local contractor.

3. You will need a 50- or 100-foot tape measure to lay out the course. Get a can of white highway marking paint. Once you have set up and driven the course and are satisfied with the layout, put a dot of paint on the pavement under each cone. Do not mess up your sponsor's pavement with a lot of paint. Try to use the layout shown in this book so that you can copy the page for each of your drivers. If you create a different layout, make a sketch and distribute a copy to each driver. The drivers need some idea of the layout before they start, since the parking lot will appear to be a confusing array of cones until they get an idea of the proper route.

4. The course shown in Figure 1-1 was laid out to fit a small parking lot. Because it loops back on itself, it is helpful to use colored flags to help define the driving route.

5. Designate one person as the officer in charge (OIC) of the course. This person assigns drivers to apparatus and starts the apparatus on the course. Designate a separate person to be the safety officer. The safety officer observes apparatus movement and monitors the course area for unauthorized persons. Both officers should have a portable radio operating on the same frequency as the apparatus. Either officer can order all apparatus movement to stop at any time. Two or three members should be familiar with the course, and one of those members should ride the officer's seat of each vehicle. Their job is to keep the driver on the correct route and to monitor the radio for orders from the OIC or safety officer. If you are training at night, ensure that all personnel are wearing clothing with reflective stripes. It is helpful if the OIC and safety officer have traffic wands on their flashlights. This will help them identify themselves as well as control the apparatus.

6. Brief the crew members of the apparatus on what to say to the public. This drill will draw attention. Crew members should be prepared to keep spectators off of the course and out of the line of motion of the training vehicles. Direct spectators to an area established by the training officer. This is a perfect opportunity to hand out fire prevention materials and to advertise your department's training regimes and professionalism.

7. Read the material in the references, and review your department's SOPs for emergency vehicle operations. Vincent Dunn's article on defensive driving and Barry McKinnon's article on siren effectiveness provide information that will help you motivate the troops.

Running the Drill

1. Set up the course prior to the drill, then brief your drivers and crews. Post a guard on the course once the cones have been set up. Cones are valuable to a lot of people for a lot of different reasons. Take precautions that your cones, or especially those that you have borrowed, do not disappear. Leave someone on guard if your company is called to respond during the drill.

2. You can record the time it takes for each driver to complete the course, but do not let your drivers focus on getting the fastest time. Concentrate on the perfect score (no cones hit) and emphasize that skill. There is a reasonable balance, however. If a driver requires 15 minutes to complete the course, then that driver needs practice.

3. Keep a log of driver training to document your department's commitment to safety. Be cautious about any notes on driver performance that you might place in the log. Remember that the document can be called into court as evidence in case of an accident.

Predrill Briefing

"This is a drill to help you improve your ability to maneuver the apparatus. We plan to do this drill regularly to establish a history of practice and build your reputation as a competent driver in case you do have an accident and are called into court. It is not a speed drill.

"The course requires skill in turning and backing the apparatus. You will be given a map of the course, and someone will be riding with you to help you find your way through it.

"_____ will be the officer in charge of the drill. He will start you on the course and monitor your performance.

"_____ will be the safety officer. He will monitor your performance and watch for spectators on the course or any other unsafe condition.

"Both officers will have radios on the same frequency as the apparatus. Drivers and riders must pay attention to the radio. If an order is given for vehicles to stop, *all vehicles must stop immediately*.

"Crew members have an active role in this drill. Your job is to keep spectators off of the course—politely. Make this a public relations opportunity. Explain to people that you are practicing skills. Explain to them how large, heavy, and cumbersome fire vehicles are and the importance of yielding the right of way.

"If people want to watch the drill, make sure that they stand in a designated area along the side of straight runs along the course, at least 15 feet back from the travel lane. Do not let anyone stand in any position directly in line with approaching or turning apparatus."

Volunteer Training Drills

Be prepared to answer the following questions:

Are drivers who hit cones allowed to drive on emergency calls?

Are drivers required to have a minimum score before they are allowed on emergencies?

Debriefing

Hold a debriefing after the drill. Find out what maneuvers were most difficult for the drivers. Find out what comments were made by the spectators.

Notes

What went right:

_____

_____

_____

_____

What went wrong:

_____

_____

_____

_____

What to do differently next time:

_____

_____

_____

_____

Figure 1-1

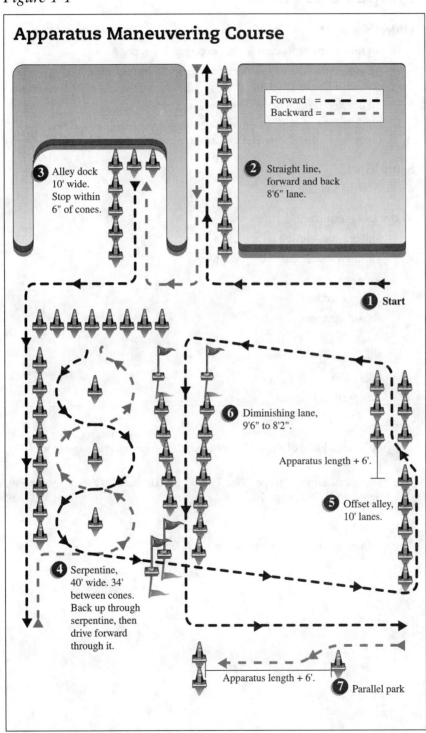

2. Ropes and Knots

Objectives

To review the procedure for inspecting utility ropes.

To ensure that the members can competently inspect rope and log the results of that inspection.

To develop proficiency in knots and rope throws.

Note that this drill is for utility rope usage only. Lifelines and rescue ropes have special care and use requirements.

Setup Time

About 15 minutes.

Materials Required

Utility ropes carried on apparatus.

Personal ropes carried by firefighters.

Two sawhorses.

Straight ladder.

Basket stretcher.

Axes.

Fans.

Pike poles.

Hose roll and nozzle.

References

"It's Just a Piece of Rope," Tom Brennan, *Fire Engineering*, February 1994, page 82.

"Personal/Utility Rope Use," George Howard, *Fire Engineering*, September 1991, page 16.

Preparation

Ensure that the proper equipment is available.

Running the Drill

1. Have the members remove the ropes from the apparatus and conduct an inspection of each one. Look for frays and cuts. Open the twisted ropes and inspect them for dirt and damage. Enter the results of the inspection in a company log. Show the date that each rope was placed in service and the date of each inspection.

2. Have the members place a ladder flat across the sawhorses. Have each member demonstrate the clove hitch and figure of eight knots on the ladder. Have each member tie these knots (1) facing the ladder, (2) facing the ladder with his eyes closed, and (3) with his back turned to the ladder.

3. Have each member demonstrate attaching a rope to raise the ladder using a figure of eight on a bight.

4. Have each member demonstrate tying a rescue knot.

5. Have each member demonstrate tying a line to raise or lower a basket stretcher.

6. Practice throwing the rope that you would use for a water rescue. Place the members at one end of the apparatus bay with a trash can for a target at the other end. Strive for both distance and accuracy.

7. Assemble the members in two relay teams at one end of the apparatus bay. Place a straight ladder, ax, two hose rolls with nozzles, a pike pole, and a fan at the other end of the bay. Place a rope beside each.

    (a) The first member ties a figure of eight on a bight and places it on the ladder for raising, then runs back and tags the second member.

    (b) The second member ties a clove hitch and half-hitch on an ax for raising, then runs back and tags the third member.

    (c) The third member ties a hitch on the hose and nozzle for raising an uncharged line, then runs back and tags the fourth member.

    (d) The fourth member ties a hitch on the hose and nozzle for raising a charged line, then runs back and tags the fifth member.

    (e) The fifth member ties a clove hitch and a half-hitch on a pike pole for raising, then runs back and tags the sixth member.

    (f) The sixth member ties a figure of eight on the fan for raising, then runs back to the starting line to stop the clock.

You can run this exercise either by telling each member his assignment in advance or by telling them when they run up to the equipment. Two teams of two members compete by having each firefighter tie three knots. Two teams of three members compete by having each firefighter tie two knots.

Volunteer Training Drills

Predrill Briefing

"This is a drill to improve proficiency with ropes and knots. We are required to keep a record of the ropes that we use and the inspections of those ropes, so start by taking the ropes off of the apparatus and checking them for rot, cuts, abrasions, or anything else that might weaken them and cause a failure.

"Then take a straight ladder, place it across a couple of sawhorses, and give everyone the opportunity to tie all of the basic knots.

"There are occasions when you might have to throw a rope to a victim to make a rescue. You probably don't get much practice in this, and there is a knack to throwing a rope accurately. See what you can do in terms of accuracy and distance so that you don't waste any time during a rescue attempt.

"Finally, we will finish up with a relay race, with each member assigned to tie a certain knot on an object. The race will be timed. It is a requirement that each knot be tied correctly."

Notes

What went right:

_____
_____
_____
_____

What went wrong:

_____
_____
_____
_____

What to do differently next time:

_____
_____
_____
_____

3. Roof Operations

Objective
To review safe operating procedures on roofs.

Setup Time
An afternoon to choose buildings on which to raise ladders and get owner permission, then:

15 minutes if you have the video (see below); two hours if you don't.

Materials Required
VCR/TV

The Fire Engineering video *Peaked Roof Collapse*.

Copies of the pretest.

Pencils.

References
"Beware of Hidden Dangers in Wood Truss Fires," John J. Novak, *Fire Engineering*, August 1997, page 153.

"Roof Operations: Lightweight Trusses and Variations," Bob Pressler, *Fire Engineering*, June 1997, page 49.

"The Roof," Tom Brennan, *Fire Engineering*, December 1996, page 94.

"Beware the Roof," Ray Downey, *Fire Engineering*, May 1996, page 63.

"Operations on Peaked-Roof Structures," Bob Pressler, *Fire Engineering*, July 1995, page 18.

"Safety During Roof Operations," Gene Carlson, *Fire Engineering*, November 1994, page 12.

"Fort Worth Collapse: Our Continuing Lessons," Jeffrey Harwell, *Fire Engineering*, January 1993, page 33.

"Get the Roof," Matthew Murtagh, *Fire Engineering*, March 1991, page 16.

"Peaked-Roof Safety," John Mittendorf, *Fire Engineering*, July 1990, page 28.

"The Peaked Roof," Vincent Dunn, *Fire Engineering*, March 1990, page 36.

Volunteer Training Drills

Preparation

1. Make copies of the quiz and bring pencils for more than the number of persons expected at the drill. Take the quiz yourself and be sure you know the answers. If you have the video, set up a VCR. If you do not have the video, set up a chalkboard and prepare a presentation following the attached outline. The *Fire Engineering* articles noted above (especially the one by Vincent Dunn) will help you prepare your presentation.

2. In good weather, consider taking the apparatus out after the classroom session and develop proficiency in placing firefighters and tools on roofs with high parapets, open shafts, and other hazards. If you plan on taking the company to a specific building, let the owners know that you are coming. Tell them what you plan to do and why. Get their permission to use the building. If they object to having people on the roof, ask whether you can raise ladders and at least look at the roof from the ladders.

Running the Drill

1. Give each member the quiz. Allow them about 10 minutes.

2. Afterward, allow them to correct their own quizzes as they find the answers in the video or your presentation.

3. After the video or presentation, go over the answers to the quiz. Open a discussion of problem roofs, particularly those with firefighter safety hazards such as high parapets, open shafts, bowstring trusses, and slate surfaces. Discuss how to overcome those hazards.

4. If time and weather permit, go to problem roofs, raise ladders, and give all of the members a good look at the hazards.

Predrill Briefing

"The purpose of this drill is to review safe operating procedures for roofs. The specific objectives are to review the procedures for getting on and off roofs, for moving around on sloped roofs, for evaluating the stability of a roof structure, and to fine-tune emergency procedures in case you get trapped on a roof."

Debriefing

Hold a general discussion of the problem roofs in your response area. What buildings did your firefighters list as problems on the last question of the test? Why did they select those particular buildings? Which ones are weak? Which of them will be very slippery when wet? What tactics will you use to operate safely on those roofs?

Notes

What went right:

_____
_____
_____
_____

What went wrong:

_____
_____
_____
_____

What to do differently next time:

_____
_____
_____
_____

Figure 3-1

> ## Roof Operations Quiz
>
> 1. To operate safely on a peaked roof, a firefighter should be familiar with which elements of roof construction?
>   (a) The roof rafter system.
>   (b) The roof deck below the shingles.
>   (c) The slope of the roof.
>   (d) The shingle or roof covering.
>   (e) All of the above.
>
> 2. Which of the following will safeguard a firefighter against a peaked-roof collapse?
>   (a) Walking on the roof ridge.
>   (b) Walking on the rungs of a roof ladder.
>   (c) Walking flat-footed, bending the legs at the knees.
>   (d) Operating from an aerial ladder or platform.
>
> 3. What length roof ladders should be carried on the first-due engine?
>
> 4. What length roof ladders should be carried on the first-due truck?
>
> 5. What is the minimum length of a roof ladder for it to be effective?
>
> 6. What is the steepest roof slope that you might safely walk on without a roof ladder?
>
> 7. Describe three safety procedures when walking on a sloped roof.
>
> 8. You are operating on a peaked roof and are blinded by smoke. You cannot see, and the smoke is increasing. What should you do?
>
> 9. Describe how to climb safely from a ground ladder to a roof ladder.
>
> 10. Describe how you would vent the roof of a typical single-family dwelling.
>
> 11. Describe the hazards of a building with a slate roof.
>
> 12. Describe what you should look for and what you should do if ordered to the roof of a commercial structure during a working fire.
>
> 13. You are ordered to the roof of a drugstore. You position a ladder in front of the building and climb to the top. It is dark, and the visibility is poor. What should you do before stepping off the ladder?
>
> 14. List the buildings in the first-due area that present special roof-access or safety problems.

Figure 3-2

QUIZ ANSWERS

1. e

2. d

3. Answer depends on local apparatus.

4. Answer depends on local apparatus.

5. A roof ladder should extend from the peak of the roof past the eaves, over the edge.

6. 30 degrees.

7. (a) Walk with a flat-footed step instead of the normal heel-toe step.
   (b) Walk with your knees bent.
   (c) Walk at an angle down the slope of a peaked roof. Do not walk directly down the roof.

8. Crawl up the roof to the ridgepole and move along the ridgepole to safety.

9. The ground ladder should extend four to five rungs above the roof. Hook any tools to the roof ladder. Step only on a rung of the roof ladder that is supported by the roof. Do not place any weight on the portion of the roof ladder that is overhanging the roof.

10. Over the fire, as high as possible, near the ridgepole.

11. Poor footing, very slippery when wet. Tiles can break loose, causing a roof firefighter to lose his balance or injuring firefighters on the ground.

12. (a) Size up the roof for immediate hazards.
    (b) Size up escape routes.
    (c) Check for trapped occupants at windows, balconies, and fire escapes.
    (d) Communicate with Command.
    (e) Locate the fire.
    (f) Ventilate using roof scuttles, skylights, and doors.
    (g) Remove window glass for a small, single-room fire on the top floor.
    (h) Cut a ventilation hole for a multiroom or cockloft fire.

13. Use a pike pole to check the height of the parapet.

14. The list of buildings should include balloon construction, truss roofs, slate or tile roofs, very steep roofs, and sheet metal roofs.

Figure 3-3

Peaked Roof Collapse—Lecture Outline

Common Types of Peaked Roofs:
1. Gable.
2. Hip.
3. Gambrel.
4. Shed.
5. Combination.

Most fires occur in residential buildings, the majority of which have peaked roofs.

Preplanning
1. Beware of overconfidence.
2. Know as much as possible about peaked-roof construction.
3. Be familiar with local construction.
4. Conduct inspection visits. Make sure that the information gets passed to all members and that it is incorporated in preplans.
5. If you don't know the construction method when responding, consider all of the possible hazards.

Types of Peaked Roof Support
1. Timber trusses—most often found in commercial buildings and spaced up to 20 feet apart.
2. Plank and beam—typically used for barns.
3. Rafters—solid-wood beams spaced 16 to 24 inches on center.
4. Pre-1960—most of these buildings have a solid-beam roof supported by solid-beam rafters. (Where are these buildings in the first-due area?)
5. Post-1960—many of these buildings have peaked roofs supported by lightweight wood trusses. They pose a higher potential for collapse. Where are these buildings in the first-due area?

Roof Decks

1. Construction—thin tongue-and-groove wood boards; otherwise, plywood or composition board nailed to rafters or trusses.
2. Covering—tar paper covered by shingles. The shingles may be slate, tile, asphalt, wood, asbestos, or sheet metal.
3. Shingle hazards:
   (a) Slate or tile shingles are razor sharp and can be deadly when dislodged and sent flying through the air.
   (b) There may be no roof decking (missing or rotted out) beneath the shingles. Such an area may collapse when you step on it.

Reasons for Operating on a Peaked Roof

1. To extinguish a shingle fire.
2. To extinguish a chimney fire.
3. To vent an attic fire or vent the building vertically.
4. To overhaul a smoldering roof deck fire.

Precautions to Prevent Falls

1. Be extra careful. More firefighters are injured in falls from peaked roofs than from roof collapses.
2. Bring along only the tools you intend to use. Extra tools make it harder to climb, they throw you off balance, and they leave you with no free hand to gain control. Falling tools are hazardous to other members.
3. Roofs are frequently too steep for safe walking. A roof angle of 30 to 45 degrees requires the use of a roof ladder.
4. Beware of a slippery surface caused by water, snow, moss, mildew, sap, or heated tar.
5. Stay near the roof ridge when walking. It is the one sure handhold. Chimneys, vent pipes, and TV antennas may collapse.
6. Walk flat-footed, with your knees bent.
7. If you must walk near the edge of a roof, walk diagonally across the roof, not straight down to the edge. Forward momentum may cause you to fall.
8. Never run on a roof. You may not be able to stop quickly without falling.

Causes of Peaked Roof Collapse

1. Failure of one of the primary members.
   (a) Bearing walls.
   (b) Hip rafters.
   (c) Ridge rafters.
2. Excessive weight of firefighters.
3. Rotted roof decking. Look for rot wherever snow, ice, and rain can accumulate—near the eaves, where the slope of the roof changes, or where a sloping roof abuts a vertical surface.

Collapse Precautions

1. Use a roof ladder if you suspect the roof deck to be weak or poorly supported.

    (a) Support the ladder with the ridgepole.

    (b) Walk on the rungs.

2. Use an aerial ladder or platform if there is a possibility of roof rafter collapse.
3. Keep operating time to a minimum below a peaked roof or in the area around the fire building, thereby avoiding injury when the roof deck becomes weakened by fire, master streams, or sudden changes in temperature.
4. Keep personnel clear of areas under or in the path of powerful master streams.
5. Do not operate on or under the roof if the fire has involved the supporting members of a lightweight truss roof.

Additional Dangers

1. Chimney collapse.
2. Falling through skylights.
3. Porch, dormer, or cornice collapse.

Questions for Discussion

1. What are the department's safe operating procedures for peaked roofs?
2. What peaked-roof buildings in the first-due area are too dangerous to operate on during a fire?
3. What type of fire in a peaked-roof residential building does not require roof ventilation?
4. If you vented a roof and a sudden shift in the wind left you blinded by smoke, what would be the safest action? (Crawl up the roof to the ridgepole, then along the ridgepole to safety.)

4. Ground Ladders

Objectives
To gain proficiency in carrying, raising, climbing, and lowering ground ladders. To improve inspection and maintenance skills.

Setup Time
About one hour.

Materials Required
Ground ladders carried on apparatus.

Checklist for ladder inspection (distribute copies to all members).

Checklist for ladder handling (copies to all members).

Soap, brushes.

SCBA.

References
"Operational Variations for Ground Ladders," John W. Mittendorf, *Fire Engineering*, October 1997, page 106.

"Ground Ladder Placement," John W. Mittendorf, *Fire Engineering*, September 1997, page 59.

"Portable Ladders and More," Tom Brennan, *Fire Engineering*, March 1993, page 190.

"I'll Just Throw Up a Ladder and Get Them!", Tom Brennan, *Fire Engineering*, June 1992, page 150.

NFPA 1932, *Use, Maintenance, and Service Testing of Fire Department Ground Ladders*.

Preparation
1. Select a site where an apparatus can be placed near a building that can be laddered. A multilevel building is ideal. Obtain permission from the building owner to ladder the building for drill purposes.
2. Review the chapter on ladders in *Essentials of Fire Fighting*.
3. Make copies of the inspection and evaluation checklists to distribute to the members.
4. Obtain a mild detergent, scrub brushes, and waterproof grease (the grease for pawls and rollers per manufacturer's recommendations).
5. Assign two members to be safety officers and evaluators.

Volunteer Training Drills

Running the Drill

1. Have crews remove the ladders from the apparatus and place them on the ground.
2. Conduct an inspection of each ladder using the checklist supplied in this book.
3. Have the members estimate the heights of the roof and windows, then select the appropriate ladders.
4. Place two ladders against the building, either to the roof or to separate windows on the same floor. Make sure that the ladder tips are tied in and that a firefighter foots each ladder.
5. Have each firefighter climb one ladder and return down the second ladder. Each firefighter must perform this exercise three times:

    (a) once wearing turnout gear.

    (b) once wearing turnout gear and SCBA.

    (c) once wearing turnout gear and SCBA and carrying a tool.

6. Assign members to place ladders in position for:

    (a) venting a window.

    (b) entering a window for search and rescue.

    (c) lowering a victim from a window by rope.

7. Practice climbing and leg-lock techniques for working with tools and operating hoselines.
8. Go through the procedure for lowering a victim from a window using a rope placed through the rungs of a ground ladder (see Chapter 5 of IFSTA's *Essentials of Fire Fighting*). Use both a basket stretcher and a rescue knot to lower a simulated victim. Raises and lowers from heights using live victims are not recommended except during actual emergencies.
9. At the completion of the session, wash all of the ladders with mild soap. Apply waterproof grease to moving parts. Return all of the ladders to the apparatus.

Predrill Briefing

"This is a drill to improve the handling of ground ladders as well as climbing technique. We will go through evolutions of selecting the proper ladders, raising and climbing ladders, placing ladders for venting and rescue, and inspecting and cleaning the ladders.

"Each of you has been given a checklist that contains key points in the handling of ladders. Two members will be acting as safety officers and evaluating performance based on the checklist. Take a look at the checklist before you go outside and see whether you have any questions about it."

Debriefing

Have the evaluators review the drill and the performance of the members. Make sure that all of the criticism is presented positively. Emphasize that your goal is to accomplish good ladder handling skills without injuries.

Notes

What went right:

_____
_____
_____
_____

What went wrong:

_____
_____
_____
_____

What to do differently next time:

_____
_____
_____
_____

Figure 4-1

Checklist for Ladder Inspections

Per NFPA 1932, ladders should be inspected monthly and after each use. Verify that:

1. The ladder is neither burned nor discolored.
2. Heat indicators placed on the ladder are within the expiration date and do not indicate overheating.
3. The ladder is neither twisted nor warped.
4. Wooden ladder beams have no cracks, splintering, breaks, checks, wavy conditions, or deformation. The beams have no dark streaks in the wood. The bolts are tight without crushing the wood.
5. Aluminum ladder beams do not have any weld cracks. This requires close inspection. Very small cracks in aluminum can spread instantly under load, causing failure of the ladder.
6. The sliding areas have no galling.
7. The rungs of aluminum ladders are not bent and have no punctures or worn serrations.
8. Wooden ladder rungs do not have any cracks, splintering, breaks, checks, wavy conditions, or deformation.
9. All of the ladder rungs are tight and snug.
10. The halyard pulley turns freely and the pulley strap does not have any cracks.
11. The strands of the halyard show no cuts, fraying, kinks, or rot.
12. The dogs operate easily and do not stick. Inspect them for cracks.
13. The ladder raises and lowers smoothly.
14. The hooks of roof ladders are sharp.
15. The butt spurs do not have excessive wear or other defects.

Figure 4-2

Checklist for Ladder Skills

Carrying a Ground Ladder
1. Lifts with knees bent to prevent back strain.
2. Uses correct ladder carry.
3. Brings the ladder heel-first toward building.

Placement for Ventilation
1. Checks for overhead wires and obstructions before raising.
2. Places ladder windward of the opening, three to four rungs above the sill.
3. Raises the ladder smoothly.
4. Checks to be sure the dogs are set.
5. Checks for correct angle.
6. Ties off the halyard with a clove hitch and a half-hitch.
7. Heels the ladder. Firefighter heeling the ladder does not look up.
8. Firefighter on ladder locks in (leg or belt).
9. Protective gear on before ventilating.

Placement for Entry—Wide Window
1. Checks for overhead wires and obstructions before raising.
2. Places the ladder to one side of the opening.
3. Places the ladder with three rungs through the opening.
4. Checks for correct angle.
5. Checks dogs.
6. Ties off the halyard with a clove hitch and a half-hitch.
7. Heels the ladder. Firefighter heeling the ladder does not look up.
8. Protective gear on before ventilating.
9. Proper spacing of people on the ladder.
10. Ladder is tied in.
11. Raises hose by rope or pike pole.

Placement for Narrow Window or Rescue
1. Checks for overhead wires and obstructions before raising.
2. Places the ladder in the center of the opening.
3. Places the tip of the ladder at the sill.
4. Checks for correct angle.
5. Checks dogs.
6. Ties off the halyard with a clove hitch and a half-hitch.
7. Heels the ladder. Firefighter heeling the ladder does not look up.
8. Protective gear on before entry.
9. Proper spacing of people on the ladder.
10. Ladder is tied in.
11. Raises hose by rope or pike pole.
12. Two firefighters enter for rescue.
13. Victim is placed with his face toward the ladder.
14. Firefighter positions his hand to protect the victim's head.

5. Rescue by Ladder

Objective
To gain proficiency in rapidly placing a ground ladder for rescue and successfully removing a victim.

Setup Time
About one hour.

Materials Required
Ground ladders carried on apparatus.

Stopwatch.

Rescue dummy.

SCBA.

References
"Methods for Safe Aerial Device Operations," John Mittendorf, *Fire Engineering*, June 1996, page 79.

"Victim Removal," Bob Pressler, *Fire Engineering*, November 1994, page 50.

"Portable Ladders and More," Tom Brennan, *Fire Engineering*, March 1993, page 190.

"Rescues from Windows: Warrington Court Apartments Fire, Philadelphia," Thomas Garrity, *Fire Engineering*, March 1993, page 76.

"I'll Just Throw Up a Ladder and Get Them!", Tom Brennan, *Fire Engineering*, June 1992, page 150.

"Rescue or Removal," Tom Brennan, *Fire Engineering*, January 1991, page 106.

Preparation

1. Select a site where an apparatus can be placed near a building that can be laddered and where a rescue can be simulated from a second- or third-story window. Obtain permission from the building owner to use the building for the drill.
2. Review the chapter on ladders in IFSTA's *Essentials of Fire Fighting*. Conduct Drill 4, Ground Ladders, before initiating this drill.
3. If no rescue dummy is available, construct one using hose, duct tape, and a pair of coveralls as shown in Figure 5-1. Place the dummy just inside the window, leaning against the wall.
4. Obtain a doll about the size of a CPR baby. Place it somewhere else in the room.
5. Assign two members to be safety officers and evaluators.

Running the Drill

1. Have an engine or truck crew "respond" to the drill side of the building.
2. Tell the officer that a victim was seen at the window and is believed to be inside and unconscious. Fire is below the victim and extending rapidly, and the interior stairs are involved. Start the stopwatch.
3. The officer should immediately order a ladder to the window for the rescue. The ladder should be raised to the sill properly and securely footed.
4. The first member up the ladder should move the victim back into the room and enter. If sufficient staff is available, two members should enter the room. All firefighters should be in full protective clothing and be wearing masks when they enter.
5. The interior firefighters assist in placing the victim on the ladder. They should then conduct a search before exiting. They should not have to be coached to make the search.
6. Note the time that the first victim is brought down. Stop the watch when they bring the baby down.
7. Record the air remaining in the firefighters' SCBA.
8. Critique the exercise using the ladder skills checklist from Drill 4.

Predrill Briefing

"This is a drill to develop skills for rescue by ladder. You will be given a scenario when you pull up to the scene. The evolution will be timed, starting when you are given instructions and ending when the rescue has been completed. You will be required to use SCBA when you enter the building. Two evaluators will be monitoring the time, technique, safety, and air usage. See how long it takes to accomplish a rescue, and work on ways to reduce that time.

"Your ability to reach and rescue a victim is an absolute measure of your professionalism. Rescue is a difficult task to accomplish. Every firefighter needs the chance to practice these skills. There will be no second chances for a victim in an actual fire.

"Are there any questions before you begin this exercise?"

Debriefing

Have the evaluators review the drill and the performance of the members. Make sure that all criticism is presented positively. Emphasize that your goal is to accomplish the rescue.

Notes

What went right:

___

What went wrong:

___

What to do differently next time:

___

Volunteer Training Drills

Figure 5-1

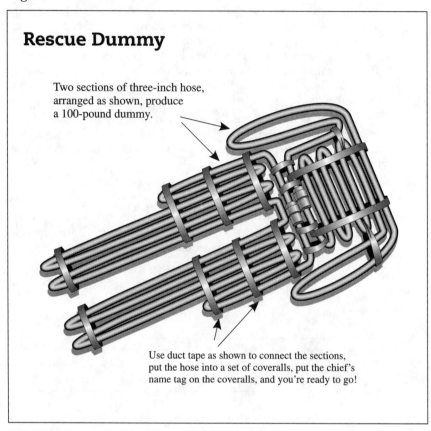

6. Preplans

Objective

To update and standardize preplan information for target hazards in the first-due area.

Setup Time

About 30 minutes if you use the enclosed Quick Access Preplan form or some other form established by your jurisdiction.

Materials Required

Copies of the preplan form.

References

"Preincident Planning and Firefighter Safety: A Success Story," Jerry Knapp, *Fire Engineering*, February 1998, page 87.

"Preincident Surveying for Highway Hazards," Eric G. Bachman, *Fire Engineering*, February, 1998, page 107.

"Preplanning Building Hazards," Francis L. Brannigan, *Fire Engineering*, August 1997, page 150.

"A 'Two-Story' House" (When platform becomes balloon construction), Ray McCormack, *Fire Engineering*, March 1996, page 66.

"Preplanning Special Industrial Hazards," Bill Gustin, *Fire Engineering*, November 1994, page 39.

"Preplanning Building Hazards," Frank Brannigan, *Fire Engineering*, November 1994, page 65.

"Pressure-Regulating Devices in Water-Based Fire Protection Systems," Walter Damon, *Fire Engineering*, November 1993, page 51.

"Prefire Planning: Anything But Routine," Roger Feagley, *Fire Engineering*, March 1992, page 119.

"Preplanning for Apartment Complex Response," John Burkush and Jim Forrest, *Fire Engineering*, January 1992, page 67.

Preparation

1. Prepare a list of target hazards.
2. Depending on the number of personnel and vehicles available, select the most serious hazards in the greatest need of updating.

Running the Drill

The drill is self-running, following the predrill briefing. Members are assigned to various target hazards to update existing preplan forms and to create quick-plan forms.

Predrill Briefing

Introduce the idea of a quick plan—a one-page summary of information critical to tactical decision making.

Emphasize that each member may one day find himself in the officer's seat, where he will be depending on that quick-plan information to help him with the first key decisions.

Emphasize that ordinary citizens know their way around major public buildings and that they expect firefighters to know their way around also.

Go through each part of the quick-plan form. Review the calculation to estimate the fire flow needed for various percentages of fire involvement.

Instruct each member to contact the owner of the building that they are preplanning before they start looking around. They should show the owner the type of information that they are trying to gather and explain why it is important.

Pay special attention to roof construction and support systems, as well as the location of fire department connections and the closest hydrant or water source.

Debriefing

Hold a general session to answer questions about the type or format of information to be collected.

The follow-up to this drill is a session in which each member or team preparing a preplan presents its findings to the other members. This is an excellent backup drill when your planned drill falls apart at the last minute.

DRILL OUTLINES

Notes

What went right:

_____
_____
_____
_____

What went wrong:

_____
_____
_____
_____

What to do differently next time:

_____
_____
_____

Volunteer Training Drills

Figure 6-1

Quick Access Preplan

Name: _____

Address: _____

Construction: _____ Size: _____

Roof: _____

Occupancy: _____

Hazards: ☐ Life ☐ Haz Mat ☐ Collapse

% involvement	10%	25%	50%	75%	100%
Fire flow per floor					
gpm required					
Number of engines required					
Number of alarms required					

Staging area _____

Water Supply
Location and capacity (include secondary supply)

Sprinkler connection: Side___   Protection: Full/Partial

Standpipe connection: Side___   Class: 1 2 3

Access
Means of entry: _____

Strategy
_____

Problems
_____

***RESCUE-EXPOSURES-VENTILATE-CONFINE-EXTINGUISH
CHECK FOR EXTENSION***

☐ Primary search   ☐ Backup line   ☐ Secondary water supply

7. Physical Fitness I

Objectives

To give the members a chance to evaluate their own physical fitness and provide a goal-setting worksheet. Consider running this drill once every three to six months so that the members can measure their progress.

Setup Time

About an hour to construct the flexibility measurement board and about 30 minutes to make copies of the evaluation pages.

Materials Required

Copies of the self-evaluation worksheet and fitness plan.
The Fire Engineering video *Stamina*.
Cloth tape measure.
Standard tape measure.
Flexibility measurement board.
Body fat calipers.

Support Required

As a minimum, have an EMS crew available. Your department may want to require a physician's release for each member. A certified fitness instructor, if available, should conduct this drill.

References

"Assessing Firefighter Aerobic Capacity: The Rockport Field Test," John Lecuyer, *Fire Engineering,* February 1998, page 49.

"Endurance Training for Firefighters," Frank Fire Jr., *Fire Engineering*, April 1994, page 14.

"Strength Training for Firefighters," Frank Fire Jr., *Fire Engineering*, April 1993, page 45.

"Training Women for the P.A.T.," James Bird, *Fire Engineering*, March 1991, page 87.

"Physical Fitness Coordinator's Manual for Fire Departments," U.S. Fire Administration, FEMA, 1990.

"Health and Physical Readiness Standards," U.S. Navy, 1990.

NFPA 1500, *Fire Department Occupational Safety and Health Program.*

NFPA 1582, *Medical Requirements for Fire Fighters.*

Preparation

1. You will need an open, carpeted area available for workouts.
2. Advise the members to bring running shoes and workout clothes to this drill.
3. Make a copy of the Health Risk Evaluation Form (Figure 7-1), Stretching Exercises (Figure 7-2), U.S. Navy Physical Readiness Standards (Figure 7-3), and the Personal Fitness Goals and Progress Chart (Figure 7-4) for each member.

Volunteer Training Drills

4. Make a board for measuring flexibility as shown in Figure 7-5.
5. Purchase a body fat caliper--about $20 at a discount sports store.
6. If you have both male and female members, you will need both a male and a female training officer.
7. Identify a 1 1/2-mile flat course, or make arrangements to use an outdoor high school track.

Safety Notes

1. Your department may want to require a physician's release for each member. A certified fitness instructor, if available, should conduct this drill. If you are not an EMS company, arrange to have an EMS unit at the company during this drill. Even healthy-looking individuals can have hidden health problems. You owe it to your members to be prepared for an unexpected emergency.
2. Any member who has any difficulty with any exercise must be taken out of the drill immediately.
3. Ensure that the members are dressed appropriately for the weather. Have drinking water available, and make sure that the members drink water prior to and following each exercise to avoid dehydration, even in cool weather.

Running the Drill

1. This drill has three parts:
   (a) Fill out a Health Risk Evaluation Form (Figure 7-1).
   (b) Perform a series of exercises.
   (c) Prepare a Personal Fitness Goals and Progress Chart.
2. First, the members fill out the Health Risk Evaluation Form. This form is for the individual members' personal use. Be sure to record each member's resting pulse and blood pressure before starting any stretching exercises.
3. Pass out a copy of the list of stretches. Have each member go through the stretching exercise while waiting his turn for body fat and flexibility measurements.
4. Instruct the members to stop any stretching or exercise that causes discomfort. Remind the members that continuing to push through an exercise because of peer pressure can be dangerous, possibly fatal.
5. Using the attached form, take the members' measurements using the body fat caliper. A female training officer is to take the measurements of the female members, and a male training officer is to take the measurements of the male members. No exceptions. If there is only one female member, give her the option of having the measurements taken by a family member at home. Measure body fat according to the instructions provided with the calipers. Record the body fat measurement on the member's Health Risk Evaluation Form.
6. Each member should sit on the floor with his legs straight and his feet flat against the flexibility board. Instruct each to slowly reach forward and slide the marker block along the board. No bouncing! Record the measurements on each member's Health Risk Evaluation Form.

Drill Outlines

7. Have each member perform as many bent-knee sit-ups as possible in two minutes.

    (a) Caution the members to keep breathing and not to hold their breath during the exercise. Caution them to stop the exercise if they feel any back pain or have any difficulty breathing.

    (b) Sit-up technique: The member lies flat on a blanket or mat on the floor, knees bent, with his heels about 10 inches from his buttocks and his arms folded across his chest. A partner holds his feet to the floor.

    (c) The member curls up, touching his elbows to his thighs. Note that his arms are to be kept folded against his chest, and his feet must stay on the floor.

8. Each member should perform as many push-ups as possible in two minutes. They have the option of wearing or not wearing shoes, but their feet may not be placed against a wall or other fixed object. They should begin in the up position. The members may rest only in the up position, not down on the floor.

9. Have each member perform the 1½-mile run. Instruct each member to stop immediately if he experiences difficulty in breathing, chest discomfort, or leg pain. Record the time to complete the run. Instruct each member to time his own pulse rate immediately at the end of the run and at five, 10, and 15 minutes thereafter. They are to record these numbers on the Health Risk Evaluation Form. Make sure that the members walk around after the run and perform the stretching exercises to avoid injury.

10. Have each member fill out the Goals column in the Personal Fitness Goals and Progress Chart (Figure 7-4).

Predrill Briefing

"The purpose of this drill is to help you evaluate your own physical fitness and build a plan that you can follow to improve it. This is not a screening test to determine who should or should not be riding apparatus.

"The drill has three parts: the Health Risk Evaluation Form, a series of exercises, and the preparation of a plan to improve your physical fitness.

"At the beginning and end of any exercise period, it is important to warm up and cool down by stretching your muscles. You will be given a stretching chart that shows you exercises to avoid injury.

"If at any time you feel uncomfortable performing an exercise or feel as if you are pushing yourself too hard, stop and rest.

"At the end of the drill, you will be given the scoring used by the U.S. Navy to evaluate the physical fitness of its officers. You will be able to compare your results to see how you measure up. The numbers vary according to age and gender.

"You will also see a video that will give you suggestions for improving your cardiovascular fitness.

Volunteer Training Drills

Debriefing

1. Review each exercise. Emphasize the need for stretching daily.
2. Encourage the members to develop a fitness plan.
3. Show the Fire Engineering video *Stamina*.
4. If you plan to repeat the drill at regular intervals, tell the members now so that they will have an incentive to improve.
5. Consider establishing a fitness award for high-scoring individuals and a most-improved award at the end of a six-month retest period.

Notes

What went right:

_____
_____
_____
_____

What went wrong:

_____
_____
_____
_____

What to do differently next time:

_____
_____
_____
_____

Figure 7-1

Health Risk Evaluation Form

Name: _____ Date: _____

Resting pulse rate: _____ beats/minute.

Blood pressure: Resting: _____/_____ (Should be less than 140/90).

During exercise: _____/_____ (Should be less than 220/95).

(U.S. DOT Health Program recommendations)

Body fat: Your measurement: _____

	Excellent	Good	Fair	Poor
Men	Less than 16%	16-20%	21-25%	over 25%
Women	Less than 20%	20-25%	26-30%	over 30%

Flexibility: Your reach: _____ inches.

Excellent	Good	Fair	Poor
Greater than 19"	15-19"	11-15"	Less than 11"

Age: _____

	Sit-ups in 2 min.	Push-ups in 2 min.	Time: 1.5 miles
Your score			
Rating			

At completion of the 1.5-mile run, pulse rate after:

5 minutes _____   10 minutes _____   15 minutes _____

Family history of: _____

High blood pressure _____

Heart disease _____

Back injury _____

Figure 7-2

> ## Stretching Exercises
>
> **Trunk twists**—Stand with your feet shoulder-width apart. Extend your arms straight out to your sides, level with your shoulders. Twist slowly at the waist to the right. Hold for a count of 10 seconds. Return slowly and repeat to the left. Do three sets.
>
> **Arm circles**—Stand with your feet shoulder-width apart. Extend your arms straight out to the sides, level with your shoulders. Circle your arms backward, making small circles. Repeat forward. Circle your arms backward, making large circles. Repeat forward.
>
> **Toe touches**—Stand with your feet shoulder-width apart. Bend forward and touch your left toe with your right hand. Your left hand should extend straight behind your body. Return to an upright position. Repeat, touching your right toe with your left hand. This should be a slow, even stretch. Do not bounce.
>
> **Side bends**—Stand with your feet shoulder-width apart. Fold your arms behind your head. Bend slowly to the left and hold for a count of six seconds. Return to a vertical position and repeat to the right.
>
> **Ankle stretch**—Stand about 2½ to three feet from a wall, facing the wall. Lean on the wall with your arms. Bend your knees until you feel a stretching in the backs of your legs. Your feet should remain flat on the floor.
>
> **Seated stretch**—Sit with your legs straight out in front of you. Lean forward and grasp your legs. Pull with your arms until your chest touches your thighs. Your legs should remain straight.
>
> **Groin stretch**—In a seated position, bend your legs to place the soles of your feet together. Pull back on your feet, attempting to make your heels touch your buttocks. At the same time, press down on your knees very slightly.
>
> **Running in place**—Lift your feet eight to 10 inches off the floor while running in place. Run on a mat or wear running shoes to cushion the impact on your knees.

Figure 7-3

U.S. Navy Physical Readiness Standards

	17-19 Years		20-29 Years		30-39 Years		40-49 Years		50+ Years	
	Male	Female	Male	Female	Male	Female	Male	Female	Male	Female
Sit/Reach Pass/Fail	Touch Toes		Touch Toes		Touch Toes		Touch Toes		Touch Toes	
Sit-Ups (2 minutes)										
Outstanding	88	86	84	84	75	74	73	72	68	67
Excellent	72	67	68	61	54	54	48	48	45	45
Good	60	52	50	45	40	39	35	34	33	32
Satisfactory	45	40	40	33	32	27	29	24	27	22
Push-ups (2 minutes)										
Outstanding	62	36	52	29	45	23	41	22	38	21
Excellent	57	31	48	24	41	19	37	18	35	17
Good	51	24	42	17	36	11	32	11	30	10
Satisfactory	38	18	29	11	23	5	20	5	19	5
1.5 Mile Run/Walk										
Outstanding	9:00	11:30	9:15	11:30	10:00	12:00	10:15	12:15	10:45	12:45
Excellent	9:45	13:15	10:30	13:30	11:45	13:45	12:15	14:15	12:30	14:45
Good	11:00	15:00	12:00	15:00	13:45	15:30	14:30	16:15	15:15	16:45
Satisfactory	12:45	16:15	13:45	16:45	15:30	17:15	16:30	18:15	17:00	19:00

Volunteer Training Drills

Figure 7-4

Personal Fitness Goals and Progress Chart

	First test	**Goal**	**3 mos.**	**6 mos.**	**9 mos.**	**12 mos.**
Weight						
Waist size						
Chest size						
Bicep size						
Thigh size						
Resting pulse						
Resting blood pressure						
Reach in inches						
Sit-ups in 2 minutes						
Push-ups in 2 minutes						
1.5-mile run						

Figure 7-5

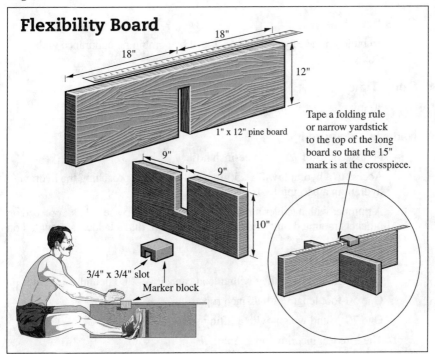

8. Physical Fitness II

Objective

To build on the Physical Fitness I drill by combining endurance with strength.

Setup Time

30 minutes to set up equipment.

Materials Required

Four sections of 2½- or three-inch hose.

Access to a hydrant with a 250-foot flat surface approaching the hydrant (a hydrant in a parking lot is ideal).

A pumper with a ladder mounted on the side. Otherwise, a ladder placed on brackets mounted on a wall at the same height that a ladder is mounted on a pumper.

One 24-foot ladder.

One single-sheave pulley with a load capacity of 300 pounds or greater.

One 50-foot length of half-inch rope.

One 70-pound weight with a lifting sling.

The Fire Engineering video *Strength*.

Advise all of the members to bring workout shoes and clothes.

Support Required

As a minimum, have an EMS crew available. Your department may want to require a physician's release for each member. A certified fitness instructor, if available, should conduct this drill.

References

"Assessing Firefighter Aerobic Capacity: The Rockport Field Test," John Lecuyer, *Fire Engineering*, February 1998, page 49.

"Endurance Training for Firefighters," Frank Fire Jr., *Fire Engineering*, April 1994, page 14.

"Strength Training for Firefighters," Frank Fire Jr., *Fire Engineering*, April 1993, page 45.

"Training Women for the P.A.T." by James Bird, *Fire Engineering*, March 1991, page 87.

"Physical Fitness Coordinator's Manual for Fire Departments," U.S. Fire Administration, FEMA, 1990.

"Health and Physical Readiness Standards," U.S. Navy, 1990.

NFPA 1500, *Fire Department Occupational Safety and Health Program*.

NFPA 1582, *Medical Requirements for Fire Fighters*.

Preparation

Set up one ladder with the pulley secured to the top of the ladder. Reeve a rope through the pulley to allow raising and lowering the 70-pound weight a distance of 15 feet. Make sure that the ladder is tied in both top and bottom.

Safety Notes

1. Your department may want to require a physician's release for each member. A certified fitness instructor, if available, should conduct this drill. If you are not an EMS company, arrange to have an EMS unit at the company during this drill. Even healthy-looking individuals can have hidden health problems. You owe it to your members to be prepared for an unexpected emergency.

2. Any member who experiences any difficulty must be taken out of the exercise immediately.

3. Make sure that the members are dressed appropriately for the weather. Have drinking water available, and make sure that the members drink water prior to and following each exercise to avoid dehydration, even in cool weather.

Running the Drill

1. Begin by going through the stretching exercises from the Physical Fitness I drill. The members should be wearing workout clothes and running shoes. Caution the members about overexertion.

2. Follow the stretching exercises with an easy five-minute jog.

3. Conduct the hose hookup exercise as follows:

    (a) Connect two lengths of 2½- or three-inch hose to one discharge of the hydrant. Measure back three lengths (150 feet) from the hydrant and draw a chalk line. Place the remaining two lengths of hose behind the chalk line.

    (b) Each member starts behind the chalk line and drags two lengths of hose to the hydrant. The member connects the hose to a hydrant discharge, then disconnects the hose already attached to the hydrant and drags it back across the chalk line.

    (c) Time starts when the member crosses the chalk line, and it ends when both lengths of the hose have been dragged back across the line.

    (d) This is harder than it sounds: Make sure each member walks around and cools down gradually after his run.

    (e) This exercise can be made a lot more interesting by dividing the members into teams and running the exercise as a relay race, timing each member in the process.

Volunteer Training Drills

4. The ladder-lift exercise consists of a single person removing and replacing a 24-foot ladder from the apparatus.

    (a) This is an excellent opportunity for back strain. Make sure there are two spotters, one on each end of the ladder, to assist in the lift if the member doing the exercise has difficulty. Instruct the spotters to assist with the ladder if the member starts to lose control or has excessive arch in his back.

    (b) This is also an excellent opportunity to ding up the paint job on the apparatus. Think about doing the exercise using a set of ladder mounts on a wall instead of using ladders mounted on apparatus.

5. Conduct the halyard pull exercise. The members must wear gloves and a helmet. Lift the 70-pound weight 15 feet and lower it three times in 30 seconds. The lifts and lowers should be under control at all times. Do not let the members slide the rope through their gloves. All of the lowers should be hand over hand.

Predrill Briefing

"The purpose of this drill is to help you evaluate your own physical fitness and add on to the personal fitness plan that you started in the Physical Fitness I drill. It is not a screening test to determine who should or should not be riding apparatus.

"The drill has three parts: a hose-pull exercise, a ladder lift, and a halyard-pull exercise.

"At the beginning and end of any exercise period, it is important to warm up and cool down by stretching the muscles. The group will first go through the stretching exercise and take a five-minute light jog before beginning the drill.

"If at any time you feel uncomfortable performing an exercise or feel as if you are pushing yourself too hard, stop and rest."

Debriefing

The drill exercises you have just gone through are part of the recruit fitness evaluation program used by the Los Angeles City Fire Department. You may not have been able to perform all three exercises to the level that you would like. The video I am going to show you will give you some techniques you can use to build strength.

Show the Fire Engineering video *Strength*.

Notes

What went right:

_____
_____
_____
_____

What went wrong:

_____
_____
_____
_____

What to do differently next time:

_____
_____
_____
_____

9. SCBA I

Objective
To review the procedures for using SCBA and to develop proficiency in its use.

Setup
About 30 minutes to arrange the search room.

Materials Required
SCBA, PASS device.

Two sections of hose plus a nozzle.

Halligan tool, broom, miscellaneous tools, and baby-size doll.

Chairs, table.

Clipboard and paper to record the starting and finishing air-bottle pressures.

Stopwatch.

References
"How to Protect Against SCBA Failures," Adam K. Theil, *Fire Engineering*, April 1998, page 97.

"Fresh Approaches to Training," Carl Welser, *Fire Engineering*, June 1996, page 12.

"SCBA Basics," Michael Terpak, *Fire Engineering*, June 1996, page 28.

"Conquering the Entanglement Hazard," Brad Bombardiere and Randy Rau, *Fire Engineering*, May 1996, page 12.

"SCBA Competence and Confidence," John Salka, *Fire Engineering*, September 1993, page 25.

Preparation
Prepare a darkened room as shown in Figure 9-1.

Running the Drill
1. Arrange the members in a circle. Leave space enough for them to don turnout gear and SCBA without injury.
2. Go over the checklist for donning SCBA (See Figure 9-2).
3. The members will be timed in this exercise as they don their gear. Stop the clock when all of the gear is in place and the members are breathing air with the PASS device turned on.
4. Have the members remove their SCBA and turnout gear, placing the equipment at their feet and turning their backs on it. Loosen the high-pressure air line connection, open the bypass and main line valves, and tighten one strap. The objective is for firefighters to learn to check their SCBA before donning it to avoid a surprise when they turn on the bottle.

5. Repeat the donning drill, timing the members.
6. Cover each member's face piece with cloth. Record the air bottle pressures. Have each member enter the search room and begin a search. When a member comes across a tool, he should identify the tool by feel. When he finds the hoseline, he should follow it to the first coupling. Have him identify the coupling and the direction to follow the hoseline out. Before exiting, have the member demonstrate the technique for removing the bottle to clear an entanglement and for sliding the bottle under his left arm to go through a narrow opening. The member completes the drill when he exits the search room. Record the air bottle pressure.

Predrill Briefing

"The purpose of this drill is to gain proficiency in searching an area in the dark while wearing SCBA. Over the years, there have been many fatalities as a result of firefighters becoming separated from the hoseline, getting lost, and running out of air.

"There will be no trick hazards in this exercise. The purpose is simply to help you build your level of skill and comfort with SCBA.

"When you enter the room, start a normal search pattern by turning to the right. Identify by feel any objects that you may find. When you find a hoseline, follow it to a coupling, then determine the direction to follow the hoseline out of the fire area. You will be asked to demonstrate removing the bottle to clear an entanglement and to demonstrate sliding the bottle under one arm to go through a narrow opening."

Debriefing

"What parts of the exercise caused difficulty?

"You all know your starting and ending pressures. The member who used the least air was _____, who used _____ psi. Compare that with your own numbers. Remember to breath in through the nose and out through the mouth. If you feel your breathing getting too rapid, slow down, get it under control, and then start moving again."

Volunteer Training Drills

Notes

What went right:

_____
_____
_____
_____

What went wrong:

_____
_____
_____
_____

What to do differently next time:

_____
_____
_____
_____

Figure 9-1

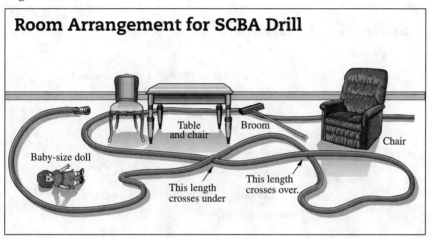

Volunteer Training Drills

Figure 9-2

SCBA Skills Checklist

Predonning

_____Checks all straps, removes twists, loosens fully.
_____All regulator valves in closed position.
_____All connections hand tight.
_____Regulator has been bled down.
_____Reads cylinder gauge out loud.
_____Opens cylinder valve slowly, listens for alarm.
_____Compares regulator gauge with cylinder gauge.
_____Uses lesser gauge reading.

Donning

_____Checks to be sure area is clear.
_____Controls and protects the regulator.

Face Piece

_____Places chin in first.
_____Places webbing or mesh over head, brushing back hair.
_____Pulls straps back, not to side. Bottom straps first!
_____Uses Nomex®/PBI® hood.
_____Fastens helmet chin strap.

Safety Check

_____Checks face piece seal (main line closed).
_____Checks exhalation valve.
_____Inhales to check regulator gauge (100 psi).
_____Conducts positive pressure check.
_____Checks bypass/purge valve for function.

10. SCBA II

Objective

To review the procedures for using SCBA and to develop proficiency in its use.

Setup Time

About 30 minutes to arrange the search room.

Materials Required

Access to a theater or auditorium that can be searched under darkened conditions.

SCBA, PASS device.

Tools, a rescue dummy, and personal rope.

References

"How to Protect Against SCBA Failures," Adam K. Thiel, *Fire Engineering*, April 1998, page 97.

"Lifeline Search—A Systematic Approach," Michael Terpak, *Fire Engineering*, January 1995, page 16.

"SCBA Competence and Confidence," John Salka, *Fire Engineering*, September 1993, page 25.

"The Search Rope," George Howard, *Fire Engineering*, January 1992, page 12.

"Personal/Utility Rope Use," George Howard, *Fire Engineering*, September 1991, page 16.

Preparation

Place a few tools or rescue dummies in the search area. Each member should have a 50-foot length of personal rope with a snap hook on each end.

Running the Drill

1. Arrange the members in a circle. Leave enough space for the firefighters to don their turnout gear and SCBA without injury.

2. Go over the checklist for donning SCBA (See Drill 9, SCBA I).

3. The members will be timed in this exercise as they don their gear. Stop the clock when all of the gear is in place and the members are breathing air with the PASS device turned on.

4. The members should be somewhat familiar with the building that they are going to search. They should be instructed that they are to bring out any victims and firefighting tools that they find.

5. Repeat the instruction that this is a search exercise. The members must not damage any property.

6. Assign teams of four to six firefighters under one officer, who will monitor the progress of the team and account for each member.
7. The members should be wearing full gear and SCBA. To gain experience in conserving air, they should be breathing from their bottles. If this presents an operational problem in terms of recharging bottles, require the members to wear SCBA and masks but not to connect the masks to the bottles.
8. The members should work their way to and then cross each aisle, staying connected by their personal ropes. Commercial structures with aisles offer many opportunities to become disoriented, lost, or entangled. The team must work together to maximize the area searched.

Predrill Briefing

"This is an exercise to become proficient in working as a team to search a large, open area with aisles, such as a theater, auditorium, or supermarket. All kinds of opportunities to get disoriented and lost exist. The only way to conduct this type of search safely is to use your personal ropes to stay linked to each other.

"Start out by hooking your ropes to your coats so that you form a chain. The lead firefighter will search down the first aisle, either to the end or to the limit of the chain formed by the team. The team will then back out and move to the next aisle. This will take some coordination among the team members, particularly in the dark. The purpose of this drill is to practice the chain technique. Any incident that would fill a supermarket, store, library, or theater with smoke would require this type of search. The drill will help you work out a system to conduct the search quickly and assure the safety of all of the members."

Debriefing

Have each team review what they did and the difficulties that they experienced.

Notes
What went right:

_____
_____
_____
_____

What went wrong:

_____
_____
_____
_____

What to do differently next time:

_____
_____
_____
_____

11. Explosive Devices

Objective
To familiarize each member with the SOPs for handling explosive devices.

Setup Time
An afternoon or evening to hold a meeting with the explosive ordnance disposal (EOD) team representative to plan the drill. If you have SOPs for responding units, review those with the EOD rep so that any needed changes can be discussed at the drill. Before the drill, allow about 30 minutes to set up an overhead projector or VCR as needed by the explosive devices team.

Materials Required
Overhead projector or VCR as needed.

References
The local jurisdiction SOPs for explosive devices.

"A Routine Call Becomes a Close Call," Larry Collins, *Fire Engineering*, October 1997, page 43.

Preparation
1. This exercise requires coordination with the agency having jurisdiction over explosive device removal. This could be either a military or a civilian agency. The major effort in preparing this drill is in arranging for a representative of that agency to make a presentation. You should specifically request the following:
   (a) A presentation on the SOPs for explosive device incidents. Ask the representative to focus on setting up a staging area, command and communications procedures, and the duties of responding units.
   (b) Ask whether there is an opportunity to stage a mock incident as part of this drill, going through a dry run of the staging and communications procedures.
   (c) Request that the representative be prepared to discuss some recent incidents and to demonstrate some of the equipment used by the explosive devices team. This will greatly improve interest in the drill.
2. Hold a meeting with the team representative to establish any special needs that he has for the drill.
3. If the drill is going to include a mock incident, prepare flyers and publicity announcements to advise the public that it is a drill. If the EOD team is a military team, invite the local law enforcement agency to participate.

Predrill Briefing
The purpose of this drill is to make all of the participants aware of the procedures to follow once a suspected explosive device has been reported. Introduce the officer of the explosive devices team and turn the drill over to him.

Drill Outlines

Debriefing

If you do not have a written SOP, use this debriefing period to prepare one. Prepare a separate page of instructions for each fire department unit that states:

(a) The procedures for responding.

(b) What to do on arrival at the staging area.

(c) The responsibilities of that unit during an incident.

(d) The communications procedures during an incident.

(e) Instructions on when and how to proceed into the hot zone in case of a detonation.

(f) The procedures for release from the staging area.

The instructions for each unit should be typed in bold letters, placed in a plastic sheet protector, and included in the preplan book for each apparatus.

Notes

What went right:

_____

_____

_____

_____

What went wrong:

_____

_____

_____

_____

What to do differently next time:

_____

_____

_____

_____

12. Arson Investigation

Objective

To familiarize each member with the indicators of and the SOPs for investigating arson.

Setup Time

An afternoon or evening to hold a meeting with the arson investigation team representative to plan the drill. If you have SOPs for responding units, review those with the arson team representative so that any needed changes can be discussed at the drill. Before the drill, allow about 30 minutes to set up an overhead projector or VCR as needed by the investigation team.

Materials Required

Overhead projector or VCR as needed.

Copies of the Arson Investigation Checklist (Figure 12-1) for each member.

References

"After the Fire's Out: Spoliation of Evidence and the Line Firefighter," Peter A. Lynch, *Fire Engineering*, January 1998, page 79.

"Investigating Automobile Fires," Lawrence Delay, *Fire Engineering*, October 1991, page 14.

"'Swiss Nanny' Trial: Fire Investigation Key to the Case," Peter R. Vallas, Peter S. Vallas, David R. Redsicker, and Arthur L. Jackson, *Fire Engineering*, December 1992, page 31.

"Arson 91," Fire Engineering, January 1991, page 44.

"Fire Investigation Handbook," U.S. Department of Commerce, NBS Handbook 134.

NFPA 921, *Guide for Fire and Explosion Investigations*.

Preparation

1. This drill requires coordination with the agency having jurisdiction over fire investigations. This could be either the fire department or the police department. The major effort in preparing this drill is in arranging for a representative of that agency to make a presentation. You should request the following:

    (a) A presentation on the SOPs for fire investigation. Ask the representative to focus on the indicators of suspicious incidents and the SOPs for firefighters.

    (b) Request that the representative be prepared to discuss some recent incidents and to demonstrate some of the equipment used in the investigation. This will greatly improve interest in the drill among the members.

2. Hold a meeting with the investigator to establish any special needs that he may have for the drill.

3. Review with the investigator the questionnaire included herein. Use it as an aid in conducting the drill. Have the firefighters answer as many of the questions as possible relative to your most recent structural fire.

4. Modify the questionnaire with any suggestions provided by the investigator. Place a copy in a plastic protective sheet in the preplan book. Have the officers of the first-arriving units use that as a reference immediately after a fire incident to identify any indicators of arson that may exist.

Predrill Briefing

Open the drill by discussing the most recent structural fire. Hand out copies of the questionnaire and ask how many members can now remember their observations as members of the first-arriving units. Introduce the arson investigator and turn the drill over to him.

Notes

What went right:

_____
_____
_____
_____

What went wrong:

_____
_____
_____
_____

What to do differently next time:

_____
_____
_____
_____

Volunteer Training Drills

Figure 12-1

Arson Investigation Checklist

First-Arriving Firefighter Observations (NBS Handbook 134)

1. General fire scene conditions on approach and arrival.
2. Weather conditions. Wind speed and direction.
3. Vehicles or persons on foot leaving the scene (description).
4. Smoke and flame colors at time of arrival.
5. Extent of involvement of structure on arrival.
6. Exact location of initial attack. Bulk of fire to left, right, or straight ahead. Fire at floor level, partway up wall, overhead, or throughout room.
7. Speed and direction of fire spread.
8. Any unusual reaction when water is first applied?
9. Any unusual odors or signs of accelerants?
    (a) Multiple fires with and without trailers?
    (b) Fire starters that did not function?
    (c) Unusual wood charring/uneven burning?
    (d) Holes in walls, floors, or ceilings made before the fire?
    (e) Candles, matches, or their remains in the debris?
10. Windows open or closed? If closed, were they locked?
11. Doors open or closed? If closed, were they locked?
12. Any evidence of forced entry, burglary, or vandalism?
13. Any evidence of tampering with fire protection systems?
14. Any contents you would not usually expect to find in this occupancy?
15. Any contents missing that you would expect to find?
16. Any evidence of tampering with utilities to produce an ignition?
17. Location of victims, if any.

13. CPR/AED

Objectives

To review the accepted standards for performing cardiopulmonary resuscitation (CPR) and using automatic external defibrillator (AED) equipment and to give the members an opportunity to build skills in a classroom and a rescue environment.

Setup Time

Check out the manikins well before the class starts.

Allow about 30 minutes to organize the equipment.

Materials Required

One or more manikins, preferably with recorders.

One or more baby manikins.

Cleaning supplies.

Replacement lungs for the manikins.

AED.

References

"Basic Life Support—Heartsaver Guide," American Heart Association.

Essentials of Emergency Care, David Limmer, Bob Elling, Michael O'Keefe, and Edward T. Dickinson, M.D., Prentice-Hall Inc., 1996.

Preparation

Schedule a certified instructor from whatever agency is approved by your jurisdiction (usually the American Red Cross or American Heart Association). Preferably, this person would be a member of the department.

Running the Drill

1. Give a short presentation, reviewing the procedures for:

    (a) One- and two-person adult CPR.

    (b) Child CPR.

    (c) Infant CPR.

    (d) Choking in adults, children, and infants.

    (e) AED operation.

2. Have each person demonstrate his skills in each of the above evolutions.

3. Have teams perform the evolution of assessing a patient and beginning CPR, loading the patient on a backboard, and moving the patient down one or two flights of stairs while maintaining CPR. Make the situation as realistic as possible. If you are out in public, post signs (and have the members be prepared to explain to the public) that this is only a drill. If at all possible, use manikins equipped with recorders. For all of its faults, the recorder does not lie!

Volunteer Training Drills

4. At the end of the practical session, have all of the members complete the recertification written test and issue new CPR cards, good for one year.

Predrill Briefing

"CPR and AED are basic skills. We are expected to be expert in CPR. Details of the technique seem to change on a regular basis. For the sake of the public, for our own reputation as professionals, and for reasons of liability, everyone is required to recertify annually."

Debriefing

Debrief the personnel, concentrating on ways to improve patient handling while maintaining CPR. Review the manikin tapes, which will show very clearly the team's performance while moving the patient.

Notes

What went right:

_____
_____
_____
_____

What went wrong:

_____
_____
_____
_____

What to do differently next time:

_____
_____
_____
_____

14. Forcible Entry—Conventional

Objectives

To review the techniques for forcible entry and to identify typical problems in the first-due area.

Setup Time

A day to tour the first-due area and take pictures.

About 30 minutes before the drill to set up the slide projector.

Materials Required

Fire Engineering's Forcible Entry Video #1, *Conventional Forcible Entry: Striking and Prying*.

A camera and slide film.

Copies of the Forcible Entry Quiz (Figure 14-1) for each member.

Pencils for more than the number of members present.

Forcible entry tools—ax, halligan, K tool, etc.

References

"The Irons," Richard A. Fritz, *Fire Engineering*, May 1998, page 93.

"Forcible Entry for Glass Storefronts," Bill Gustin, *Fire Engineering*, May 1998, page 97.

"Forcing the Tubular Dead Bolt," Bill Gustin, *Fire Engineering*, January 1996, page 89.

"Hurricane-Resistant Glass: Firefighter Resistant?" Michael Reimer, *Fire Engineering*, January 1996, page 122.

"Removing Security Bars," Bill Gustin, *Fire Engineering*, July 1995, page 40.

"The 36-Inch Pipe Wrench," Ray McCormack, *Fire Engineering*, March 1994, page 28.

"Use of the Air Chisel in Forcible Entry Operations," Bill Gustin, *Fire Engineering*, September 1994, page 55.

"Forcible Entry Tool Storage and Maintenance," Ray McCormack, *Fire Engineering*, January 1994, page 28.

"Overhead Doors and Security Bars: Forcible Entry Challenges," John Mittendorf and Lane Kemper, *Fire Engineering*, September 1993, page 63.

"Roll-Down Metal Doors," Tom Brennan, *Fire Engineering*, June 1993, page 154.

"Forcing Padlocks—An Innovation," Tom Cashin, *Fire Engineering*, February 1986, page 10.

"Forcible Entry," Tom Brennan and Paul McFadden, *Fire Engineering*, May 1984, page 36; March 1985, page 30; January 1986, page 18.

Preparation

1. Go out into the community and photograph forcible entry problems. Wear a uniform and advise the local law enforcement agency that you are taking photos for the purposes of a drill. Be sure to identify yourself to property owners and get their permission to take photos. Take an overview shot showing each building and the door of concern, as well as close-up shots of the lock mechanism. Look for:

    (a) Properties protected by guard dogs.

    (b) Properties with doors that have bars across the inside.

    (c) Properties where doors open into shafts.

    (d) Properties using locks that are especially difficult to force (Fox Police Lock or American 2000 series padlocks).

    (e) Roll-down doors, barred windows—anything that would pose an entry problem.

2. Ideally, try to locate a property that is about to be torn down so that the members can get hands-on practice with the ax and halligan.

3. Review the video and the quiz. Make sure that you know the answers ahead of time!

Running the Drill

1. Give the members the quiz. Allow them to correct their own answers as they watch the video.

2. Show the video.

3. Following the video, ask the members to identify locations in the first-due area that pose particular entry problems.

4. Using the slide show, talk about each example. Have individual members make a quick size-up and give recommendations for forcing entry.

Notes

What went right:

_____
_____
_____
_____

What went wrong:

_____
_____
_____
_____

What to do differently next time:

_____
_____
_____
_____

Figure 14-1

Forcible Entry Quiz— Conventional Techniques

1. As the forcible entry person approaching the fire building, what key decision must you make?
2. You are forcing entry into a single-family ranch-type house with light smoke showing. The door is solid and has two lock cylinders. The kitchen window is to your left as you face the door. What is the best place to force entry, and why is it the best choice?
3. What three things must you consider when forcing entry?
4. Where are most victims found?
5. You are inside a structure, conducting a search, and find a door that opens outward. Does this tell you anything special? If so, what?
6. What is the first step in forcing entry through a door?
7. How do you test for fire conditions behind a locked door?
8. How do you maintain control of a door that you are forcing?
9. List the characteristics of a good halligan bar and ax.
10. You are unable to force the locks on a door and decide to force the hinge side. Do you start with the top hinge or the bottom hinge? Why?

Figure 14-2

Quiz Answers

1. Select where to enter.
2. Select the door, because:
   (a) It provides the largest opening.
   (b) It provides an easier, safer entry and exit; the clearest path.
   (c) It is the most likely place to find victims.
   (d) Victims are more easily controlled during evacuation through a door.
   (e) The door is on a main wall, which is the best place to start an orderly search.
   (f) Heat and toxic by-products will be less threatening because you will be entering at a lower level.
3. (a) The lives and safety of civilians and firefighters.
   (b) Fire containment and extinguishment.
   (c) Damage control and ease of entry.
4. Near a door.
5. Outward-opening doors typically lead to cellar stairs, a closet, or an electrical or utility space.
6. Check to see whether the door is locked.
7. Feel the door surface with the back of your hand. Start low and move your hand up the door.
8. Hold the knob, open it slightly, and be ready to pull it closed. Do not push the door open.
9. A good halligan bar is one piece, welded or forged. An adz should have a gentle curve. A good ax weighs at least eight pounds.
10. Start with the bottom hinge. If you start with the top, the top may be pushed out, letting smoke and gas into the hallway and jamming the bottom hinge that you are trying to force.

15. Hose Handling I

Objectives

To improve hose-handling skills and to practice working as a team.

Materials Required

A pumper, water source, and attack lines.

A video camera is very useful in evaluating performance.

References

NFPA 1962, *Care, Use, and Service Testing of Fire Hose, Including Couplings and Nozzles.*

Preparation

1. Select an area where you can advance a line against a target, turn a corner, and advance against a second target (See Figure 15-1). A pair of dumpsters will work well, but don't flood the dumpsters and create a polluted runoff problem. Check to see where the water from the attack line will drain. Do not flood private property. Do not conduct this drill if freezing temperatures are expected.
2. Locate a tank trailer, LPG tank, or oil tank where you can practice controlling an LPG tank fire.

Running the Drill

1. Have the hose team advance the line against the first target, attempting to keep the stream on a specific spot. Have the team make a turn and advance on a specific spot on the second target. Then have the team follow the same path and objectives as they back the line out.
2. The objective with the LPG tank is to cool the tank with straight streams from a distance, then to approach the tank with two overlapping fog streams to reach a shutoff valve, which can be simulated. The hose teams then back out, changing from a fog stream back to a straight stream.
3. Evaluate the team's performance.

    (a) Can they control the stream and keep it on target?

    (b) Does the team keep the line straight behind the nozzle person?

    (c) Does the nozzle person maintain enough hose length to control the nozzle, or does it get pulled back?

    (d) Does the team maintain control when backing the line out?

    (e) Does water flow down both sides of the tank trailer, or is it just bouncing off the attack side?

Predrill Briefing

Introduce this drill as a skills-building session. The skill isn't as easy as it sounds. Tell the members the evaluation points that you will be using.

If you have a video camera available, tape the teams from one side of the approach to the targets. The video will show very clearly how well the teams do in keeping the stream under control.

Notes

What went right:

_____
_____
_____
_____

What went wrong:

_____
_____
_____
_____

What to do differently next time:

_____
_____
_____
_____

Volunteer Training Drills

Figure 15-1

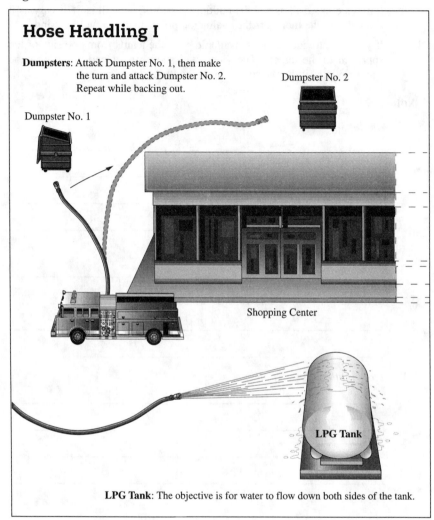

16. Pipeline Emergencies

Objective

To make all members familiar with pipeline hazards, hazard detection, and suppression methods.

Setup Time

About one hour for planning, scheduling, and organization.

15 minutes on the night of the drill to set up the drill room.

Materials Required

Visual aid equipment.

Gas meter.

References

Local SOPs for gas and pipeline emergencies.

"Gas Connectors: An Explosive Problem," Harry J. Oster, *Fire Engineering*, February 1998, page 101.

"Plastic Gas Mains: Hazards and Tactics," Leigh T. Hollins, *Fire Engineering*, December 1995, page 49.

"Night Into Day: The Edison NJ, Gas Pipeline Explosion," Albert Lamkie and David Davis, *Fire Engineering*, May 1995, page 34.

"Realistic Training for LPG Emergencies," Sanford Johnson, *Fire Engineering*, March 1993, page 83.

"Chemical Data Notebook Series #63: Methane/Natural Gas," Frank L. Fire, *Fire Engineering*, August 1991, page 101.

Preparation

1. Contact the utility that manages natural gas distribution (or propane distribution) and ask for a training session for fire suppression personnel. Every distribution company has a safety officer who provides this service.
2. Outline the information that you want the company to provide:

   (a) Describe the distribution network in your first-due area.

   (b) Describe the means of controlling the flow of gas. What types of valves are used? How are they identified and operated? Where are they located?

   (c) Describe the types and operation of the detectors the company uses to find leaks. Review the operation and maintenance of your detector.

   (d) Describe the flammability and explosive limits of the product.

   (e) Recommend suppression tactics.

   (f) Describe the company's emergency response capability—specifically, how to get help, what that help will be, and the likely response time.

   (g) Describe some incidents where things have gone well and things that have gone poorly. What lessons have been learned from these incidents?

   Recommend a time limit for the presentation that will allow for a question-and-answer period.

Volunteer Training Drills

Running the Drill

Introduce the speaker and the topic. Tell the group the information that you have requested from the speaker. That will prime them to listen for those points. Ask the speaker whether he wants to handle questions during the presentation or afterward.

Debriefing

Get feedback from the members. What was good and bad about the presentation? Be sure to send a thank-you letter to the utility company.

Notes

What went right:

_____

_____

_____

_____

What went wrong:

_____

_____

_____

_____

What to do differently next time:

_____

_____

_____

_____

DRILL OUTLINES

17. Electric Utility Emergencies

Objective

To familiarize all of the members with electrical emergencies, hazard detection, and suppression methods.

Setup Time

About one hour for planning, scheduling, and organization.

15 minutes on the night of the drill to set up the drill room.

Materials Required

Visual aid equipment.

References

Local SOPs for electric emergencies.

"Understanding Electricity and Electrical Dangers," C. Bruce Edwards, *Fire Engineering*, April 1996, page 57.

"Electrical Safety and Firefighters," George Browne, *Fire Engineering*, April 1996, page 76.

Preparation

1. Contact your local electric utility company and ask for a training session for fire suppression personnel. Every distribution company has a safety officer who provides this service.

2. Outline the information you want the company to provide:

    (a) Describe the distribution network in your first-due area.

    (b) Describe the location and layout of local distribution stations.

    (c) Describe hazards involving distribution stations and emergency access to distribution stations.

    (d) Describe the recommended practice for rescuing victims at distribution stations.

    (e) Describe the techniques used by the utility in handling hot lines.

    (f) Recommend techniques for dealing with lines down on vehicles with Priority One patients.

    (g) Describe the utility's emergency response capability—specifically, how to get help, what that help will be, and the likely response time.

    (h) Describe some incidents where things have gone well and things that have gone poorly. What lessons have been learned from these incidents?

    Recommend a time limit for the presentation that will allow for a question-and-answer period.

Volunteer Training Drills

Running the Drill
Introduce the speaker and the topic. Tell the group the information that you have requested from the speaker. That will prime them to listen for those points. Ask the speaker whether he wants to handle questions during the presentation or afterward.

Debriefing
Get feedback from the members. What was good and bad about the presentation? Be sure to send a thank-you letter to the utility company.

Notes
What went right:

_____

_____

_____

_____

What went wrong:

_____

_____

_____

_____

What to do differently next time:

_____

_____

_____

_____

18. Hose Testing

Objective

To conduct an annual hose test in accordance with NFPA standards.

Materials Required

Clipboard.

The setup checklist from this book (see Figure 18-1).

Chalk.

Marking pencil.

Test valve or gate valve.

Tags for defective hose.

References

NFPA 1962, *Care, Use, and Service Testing of Fire Hose, Including Couplings and Nozzles.*

Preparation

Review the test description in IFSTA's *Essentials of Firefighting.* Pay particular attention to the safety requirements to secure the hose, to use a test valve or gated valve to control discharge during the high-pressure phase of the test, and to maintain safety zones during the test.

Running the Drill

Follow the procedure in the checklist.

Predrill Briefing

"The purpose of this drill is to ensure that our hose is in good condition. We are required to conduct this test annually."

Volunteer Training Drills

Notes

What went right:

_____
_____
_____
_____

What went wrong:

_____
_____
_____
_____

What to do differently next time:

_____
_____
_____
_____
_____

Figure 18-1

Hose-Testing Checklist

1. Connect hose sections into 300-foot lengths.
2. Connect test valve(s) to discharge(s) opposite the pump panel.
3. Connect a test length to each test valve. Use a gated wye for two lengths.
4. Tie a hose strap from each test hose to the discharge pipe.
5. Attach a nozzle to the far end of the test section.
6. With the test valve open, fill the hose at 50 psi.
7. Raise the nozzle above the pump to bleed the air, then close the nozzle.
8. Ensure that the hose is free of kinks and twists.
9. Ensure that the couplings are tight and not leaking.
10. Mark each hose at the coupling with a pen. Place a chalk mark across the coupling.
11. Close the test valve and increase the pressure. Hold it for five minutes, maintaining a 15-foot safety zone on either side.
12. Flow a handline to prevent the pump from overheating.
13. Check for leaks at the points of hose attachment.
14. Shut down the pump and close the discharge valve.
15. Open the nozzle slowly.
16. Check the coupling for movement.
17. Tag out any hose that leaks or shows movement—$\frac{1}{16}$-inch to $\frac{1}{8}$-inch movement allowable in newly coupled hose only!

    Bring a test valve, marking pen, tags for hose that fails the test, chalk, and a copy of NFPA 1962.

19. Vehicle Extrication

Objective
To establish and practice procedures for accessing trapped victims.

Setup Time
This drill requires you to arrange for donated vehicles or to drill in a salvage yard. Once the vehicles have been procured, allow about one hour to position them, to disconnect the batteries, and to fill the gas tanks with water. Arrange to dispose of the vehicles after your drill.

Materials Required
Two vehicles that can be cut apart.

Hand and power extrication tools.

Protective clothing, including eye and ear protection.

Wood cribbing for stabilization.

Medical care equipment.

References
"Rescue Guidelines for Air Bag-Equipped Vehicles," James J. Onder, *Fire Engineering*, December 1997, page 42.

Essentials of Emergency Care, David Limmer, Bob Elling, Michael O'Keefe, and Edward T. Dickinson, M.D., Prentice-Hall, 1996.

"Auto Extrication Training Using Old and New Cars," Rob Robinson, *Fire Engineering*, December 1996, page 16.

"Vehicle Extrication Basics," Richard Marinucci and Lee Panoushek, *Fire Engineering*, May 1995, page 10.

"Advanced Vehicle Entrapment Rescue," Len Watson, Greenwave, Essex, England, 1994 (available through Holmatro Inc., Millersville, MD).

"Extricating the Brothers in Blue," George Howard, *Fire Engineering*, April 1992, page 14.

The "Carbusters" videos, American Safety Video Publishers.

Preparation
1. Advertise in your community for tax-deductible donations of old vehicles, or contact your local salvage yard.
2. Arrange with a scrap dealer to have the vehicles removed after the drill, or perform the drill at your local salvage yard.
3. Locate the vehicle on a paved surface so that glass and pieces of metal can be swept up and disposed of easily after the drill.
4. If possible, crush in the sides and/or the roof.
5. Use a rescue dummy or assign a member to simulate a victim in a vehicle. Make sure that the member wears full protective clothing, including eye and ear protection.
6. Assign crews to the engine, rescue unit, and EMS unit.

Running the Drill

Conduct the predrill briefing. If you have more than one vehicle, you may want to try one scenario using only hand tools to maintain those skills in case your power equipment fails. Evaluate the drill using the Vehicle Extrication Evaluation Points (Figure 19-1).

Predrill Briefing

"This drill will simulate a routine auto accident extrication. The incident commander will be _____. The purpose of the drill is to practice procedures to:

(a) maintain control of the scene.

(b) stabilize the vehicle.

(c) provide care to the patient.

"We will be evaluating the steps taken by the team to achieve those objectives. After the patient has been extricated, we will walk through the drill and then repeat the exercise with a second vehicle."

Debriefing

Review the exercise, using the evaluation sheet as a guide. Use this as a positive device to improve the performance of your members working together as a team.

Volunteer Training Drills

Notes

What went right:
_____
_____
_____
_____

What went wrong:
_____
_____
_____
_____

What to do differently next time:
_____
_____
_____
_____

(Figure 19-1)

Vehicle Extrication Evaluation Points

1. The incident commander maintained control of the operation:
   (a) A scene hazard survey (inner and outer circle) was conducted before approaching the vehicle. Hazards were made known to all of the rescuers.
   (b) The number of victims was determined.
   (c) Verbal contact was made to reassure the victim.
   (d) The rescuers wore proper protective equipment.
2. The scene was controlled:
   (a) A charged hoseline was placed in position and manned.
   (b) The vehicle was properly stabilized.
   (c) Tools and vehicle parts were placed so that they did not hinder the rescuers or become hazards.
   (d) A standard signal such as stop or freeze was established to halt all activity in case of a problem.
3. Patient care:
   (a) There was a clear indication that the patient was the top priority.
   (b) Communication was established with the patient.
   (c) Adequate measures were taken to prevent further injury to the patient during extrication.
4. Extrication skills:
   (a) The members worked together well to establish the steps for extrication.
   (b) The members handled the tools safely and with skill. They maintained control of the tools at all times.
   (c) The members watched out for each other's safety.
   (d) Access was provided that allowed for packaging and removal of the patient without further detriment to his condition.
   (e) Protection was placed over hazards created as a result of gaining access.
   (f) Steps were taken to disable the air bag. The members remained clear of the air-bag deployment zone.
   (g) Extrication was achieved in a reasonable time.
5. Additional evaluation points.

Volunteer Training Drills

20. Foam Hoselines

Objective

To practice the ability to set up and operate a foam line.

Materials Required

Five-gallon containers of foam, training foam, detergent, or homemade simulated foam.

Eductor, hose.

References

"Fighting Flammable Liquid Fires—A Primer," Leslie Omans, *Fire Engineering*, January 1993, page 50, and February 1993, page 50.

Preparation

Training with firefighting foam is the most effective, but it is also expensive. Training foam is less expensive. Pails of detergent (sometimes available through your state's military surplus agency) will simulate foam very effectively. You can also make a training foam solution by mixing a bottle of dishwashing detergent (not dishwasher detergent) and a box of children's bubble bath in a five-gallon container of water. Note that some foams and detergents are not environmentally friendly. Make sure that you know where any runoff is going.

Running the Drill

1. Review application rates and have the members calculate the area that can be foamed effectively with the supplies carried on first-responding units. Determine these areas for both hydrocarbon and polar solvent fuels.
2. Have the members lay out a supply line (or establish a draft), set up a foam line, and foam a designated area. Use the checklist to evaluate their performance.

Predrill Briefing

"This is a drill to evaluate your ability to set up a foam line, to ensure that the foam eductor is working properly, and to ensure that the foam is placed effectively. You may not need foam often, but when you do, everything has to work."

Debriefing

1. Go over the evaluation checklist. Note the actions that were done well and those points that need improvement.
2. Run through the calculations for foam application. Be sure that each member knows how much area can be foamed with the equipment and foam supply carried on each apparatus. Be sure that each member knows that he should wait until adequate foam supplies are on hand before beginning a foam operation.
3. Clean all of the equipment that has been used.

Notes

What went right:

_____
_____
_____
_____

What went wrong:

_____
_____
_____
_____

What to do differently next time:

_____
_____
_____
_____

Volunteer Training Drills

Figure 20-1

Foam Operations Checklist

1. The foam that is selected is appropriate for the problem. The member selects alcohol-resistant foam for polar solvent application.

2. The member sizes up the amount of foam required for the problem and does not start operations with insufficient foam.

3. The eductor is set up no more than six feet above the bottom of the foam concentrate container.

4. The eductor and nozzle are matched for foam production.

5. The foam eductor is set for the proper percentage of concentrate.

6. The pump pressure is correct.

7. The hose length, nozzle, and height of the nozzle do not create excessive back pressure.

8. The nozzle operator keeps the hose stream directed away from the fire until foam is flowing.

9. The nozzle operator does not plunge foam into the product.

10. The nozzle operator keeps his hands away from air inlets on the nozzle.

11. The nozzle operator directs the hose stream away from the fire if the foam stream turns to water.

12. The pump operator monitors the concentrate usage and keeps foam flowing to the nozzle.

21. Elevated Stream Operations

Objective

To develop proficiency in setting up a ladder pipe and other heavy streams.

Materials Required

Engine company and ladder company.

Whistle for training officers (two recommended) and safety officer.

Extra personnel to keep spectators clear.

Flares.

Two road cones.

References

Truck Company Fireground Operations, Harold Richman, The Robert J. Brady Company, 1977.

Preparation

1. The only preparation required is to identify a site with a good water source, allowing flows of 1,000 to 1,500 gpm. The ideal setup allows a 100-foot distance between the aerial ladder and a target flare, with another 100 feet to the water source. The site must allow water drainage without causing damage to property or to the environment. Do not hold the drill if freezing temperatures are expected before the site has had a chance to dry. Make sure you know where all of the water is going to go. Be sure to obtain permission to conduct the drill from the property owners.

2. This is a good drill to do at night, in a large, empty parking lot. Nighttime operations increase the challenge and reduce the number of spectators.

3. Arrange for a fill-in company, since this drill will place the participating companies out of service.

Running the Drill

This is a timed drill. Pick a location to place a flare, and spot a location for the aerial ladder about 100 feet distant.

Evolution 1: The ladder and engine company are to respond to the site. The ladder company takes a position indicated by road cones or the training officer. Start the clock. The training officer orders the ladder company to set up the ladder pipe with at least 75 feet of extension and to operate the ladder pipe from the ground with control lines to knock out the flare. The TO orders the engine company to lay dual lines from the ladder to the hydrant. The goal is to knock out the flare within three minutes of setting the brakes on the ladder truck.

Evolution 2: Place two flares about 50 feet apart and about 100 feet from the ladder's position.

The ladder and the engine arrive together. The ladder company takes a position indicated by road cones or the training officer. Start the clock. The TO orders the ladder company to set up the ladder pipe with at least 75 feet of extension and to operate the ladder pipe from the ground with control lines to knock out the right-hand flare. The TO orders the engine company to lay a single line from the ladder to the hydrant. A second TO at the hydrant orders the engine crew to pull two sections of supply line, attach a straight-tip nozzle, extend the line to the front of the ladder truck, and knock out the second flare. The goal is to knock out both flares within three minutes of setting the brakes on the ladder truck.

Evolution 3: Place two flares about 50 feet apart and about 100 feet from the ladder's position.

The ladder and engine arrive together. The ladder company takes a position indicated by road cones or the training officer. Start the clock. The TO orders the ladder company to set up the ladder pipe with at least 75 feet of extension and to operate the ladder pipe from the ground with control lines to knock out the right-hand flare. The TO orders the engine company to lay dual lines from the hydrant to the ladder. When the engine arrives at the ladder, order a single length of supply line to the ladder. Also order the deck gun to be removed from the pumper and set up to knock out the second flare with a single line from the ladder to the hydrant. The goal is to knock out both flares within four minutes of setting the brakes on the ladder truck.

Predrill Briefing

"This is a drill to develop proficiency in setting up heavy streams with an engine company and a ladder company. There will be three evolutions. They will be timed, and the training officers will let the participants know the time goal and how well they did at the end of the drill. Each of the evolutions will be different. Concentrate on working quickly but safely. If the training officers see a problem develop, they will blow a whistle. This will be a signal to stop in place."

Debriefing

What went right:

_____
_____
_____
_____

What went wrong:

_____
_____
_____
_____

What to do differently next time:

_____
_____
_____
_____

22. Engine Company Evolutions

Objective

To develop proficiency in placing lines in service.

Materials Required

Engine company.

Whistle for training officer and safety officer.

Extra personnel to keep spectators clear.

Training foam or soap solution.

References

Engine Company Fireground Operations, Harold Richman, The Robert J. Brady Company, 1975.

"Observations on the Engine Company," Andrew A. Fredericks, *Fire Engineering*, April 1998, page 83.

"Stretching and Advancing Handlines," Part 1, Andrew A. Fredericks, *Fire Engineering*, March 1997, page 56.

Advancing the Initial Attack Handline, Andrew A. Fredericks, Fire Engineering video, 1997.

Preparation

1. The only preparation required is to identify a site with a good water source, allowing flows of 1,000 to 1,500 gpm. The site must allow water drainage without causing damage to the site, adjacent property, or the environment. Do not hold the drill if freezing temperatures are expected before the site has had a chance to dry. Make sure you know where all of the water is going to go.

2. The ideal setup is a shopping center parking lot or the station's parking lot so that you can advance a line up a ladder to a roof and operate from the roof. Be sure to obtain permission from the property owners to conduct the drill. Try to allow a distance of 100 feet between the engine and a target flare, plus another 100 feet to the water source. Place traffic cones about 100 feet from the hydrant and another pair of cones about halfway between that location and the flare.

3. This is a good drill to do at night, in a large, empty parking lot. The nighttime operation increases the challenge and reduces the number of spectators.

4. Arrange for a fill-in company, since these evolutions will place the participating company out of service.

Drill Outlines

Running the Drill

Prepare 5- × 7-inch index cards with large, bold-printed instructions of the following evolutions. Have the engine company respond to the training officer, and either give the company officer the appropriate card or read the instructions to him on arrival.

Evolution 1

Lay out a supply line from the hydrant. Drive up to the first cone.

Advance the longest 1½- or 1¾-inch preconnect line to the second cone.

Charge the line from the hydrant (not the booster tank) and knock out the target flare.

Throw the 24-foot ladder to the roof of the adjoining building.

Evolution 2

Lay out a supply line from the hydrant. Drive up to the first cone.

Advance the longest 2- or 2½-inch preconnect line to the second cone.

Charge the line from the hydrant and knock out the target flare.

Throw the 24-foot ladder to the roof of the adjoining building.

Evolution 3

Lay out a supply line from the hydrant. Drive up to the first cone.

Set up the foam eductor and advance the shorter 1½-inch or 1¾-inch line from the eductor to the second cone.

Charge the line from the hydrant and knock out the target flare with foam.

Throw the 24-foot ladder to the roof of the adjoining building. Advance a second preconnect to the roof of the building.

Evolution 4

Lay out a supply line from the hydrant. Drive up to the first cone.

Throw the 24-foot ladder to the roof of the adjoining building.

Advance the longer 1½- or 1¾-inch preconnect line to the roof.

Charge the line from the hydrant and knock out the target flare.

Predrill Briefing

"This is a drill to develop proficiency in placing hoselines in service. There will be four evolutions. They will be timed, and the training officers will let the participants know the time goal and how well they did at the end of the drill. Each of the evolutions will be different. Concentrate on working quickly but safely. Each member is to be wearing full gear and SCBA but not breathing from the bottle. If the training officers see a problem develop, they will blow a whistle. This will be a signal to stop in place."

Volunteer Training Drills

Notes

What went right:

_____
_____
_____
_____

What went wrong:

_____
_____
_____
_____

What to do differently next time:

_____
_____
_____
_____

23. Residential Fires SOP—Combined Evolutions

Objective

To develop proficiency in attacking residential fires.

This is designed to be a combined engine-ladder-EMS exercise. The emphasis is on ladder crew responsibilities. However, engine crews should know all parts of the SOP in the event that the ladder company is delayed during an actual response. Each member should become familiar with the SOP for dwelling fires and run through practice evolutions.

Setup Time

About two hours of advanced preparations if you have an SOP. If you do not have an SOP, the one described in this book may be used or modified to suit your department.

Materials Required

Engine, ladder, and EMS companies.

Copies of one-page flyers to distribute throughout the neighborhoods affected.

Copies of the SOP duties for each member.

A handout on ladder crew duties for each member.

A barrel that can be placed outside with a wood fire inside.

References

Department SOP for residential fires.

Fire Officer's Handbook of Tactics, Second Edition, John Norman, Fire Engineering, 1998.

"The Private-Dwelling Fire, Bob Pressler, *Fire Engineering*, February 1998, page 22.

"Modified Frontal Attack and the Attached Garage Fire," Michael E. Haberski, *Fire Engineering*, September 1997, page 72.

"Operations in Heavily Protected Homes," Ray McCormack, *Fire Engineering*, July 1997, page 63.

"Ladder Company Operations—Private Dwellings," FDNY, March 1992.

Preparation

1. Arrange with three or four members to use their houses as drill sites for practicing the SOP.
2. Copy and distribute flyers advising neighbors of the nature of the drill. A sample flyer is provided in Figure 23-1.
3. Arrange for a member to simulate a victim at each site. At one site, provide the victim with a length of electrical cable and have the victim position himself with one end of the wire under the siding near (but clear of) an outside outlet. The victim should be face down, holding the other end of the wire in his hand.

Running the Drill

1. Provide a 30-minute introduction of the SOP. Give each member a handout that describes the duties of each person. A sample SOP and presentation are provided below.

2. Run a practice evolution on a single-story ranch-type house. A trash fire will be burning behind the house on Side 3.

    (a) Have the engine arrive and its least experienced members pull a line to extinguish the fire.

    (b) Have the ladder crew walk through their assigned duties. (Do NOT damage the building or shut off the utilities.)

    (c) Victims will be found around the house. Evaluate search effectiveness and the EMS response to and survey of victims.

3. The engine and ladder companies swap members. They respond to a two-story house and walk through the SOP again. Discuss the best entry point to the second floor. Have the driver and vent person place the ladder. Do the walkaround of the structure. Point out that the easy way to locate the utility controls is to find where the power line enters the house. Remind the members to carry pike poles and ladders carefully so as not to contact power lines. Make sure that everyone knows how to shut off the gas.

4. Repeat the drill a third time on a third structure. This time, place the victim with the simulated electric wire. Station an officer to observe the evolution and declare each rescuer who touches the victim to be an additional casualty. The victim cannot be touched until the ladder company has located the electric shutoff for the house. (Do NOT shut off the power to the house.)

Predrill Briefing

Understand that fire departments are no longer protected from lawsuits for damages if they do not perform according to recognized standards.

Emphasize that certain activities must take place and that each member must carry out his assigned task.

The SOP to be introduced in this drill is for residences. A separate SOP for commercial buildings will be presented and practiced in another drill.

The SOP is not optional. The members are to follow the SOP unless the officer tells them otherwise.

Purpose

To establish standard operations for company members to:

(a) Improve the chances for successful rescues by increasing the speed of essential operations.

(b) Increase the safety of the members by ensuring that essential jobs are accomplished and by minimizing freelancing.

(c) Minimize confusion by reducing the need for routine orders from the officers.

Background

Nationally, 70 percent of all fire deaths occur in residences. (Add the statistics for your own jurisdiction.) Special fire problems associated with residences are:

(a) Older dwellings may be of balloon construction. Expect rapid extension of fire to the attic areas.

(b) Homes of 1960s and '70s vintage have fire-resistant wallboard and generally tight construction. Anticipate high levels of heat and significant fire development before the fire is discovered.

(c) Homes built since the 1980s have lightweight truss roof construction using expanded metal fasteners. These roofs are subject to early collapse. No one should be on the roof if there is evidence of fire in the attic space.

(d) The majority of homes have central heat and air conditioning systems that are in operation all year. Anticipate rapid spread of smoke and gases to sleeping areas.

(e) Many homes are being renovated with aluminum siding. Fire involving the electrical service can charge the siding and anyone touching it.

The primary duties of the first-arriving crews are:

(a) to conduct obvious rescues (victims at windows).

(b) to force entry, then to place lines to protect the victims, stairwell, and uninvolved portions of the structure.

(c) to vent and search the fire floor and the floor above.

Concept of Operations

1. Assumptions:

    (a) That there is a light-to-medium fire in an occupied single-family residence. The structure is a wood-frame, two-story home with a peaked roof. Standard operations are based on this situation.

    (b) The engine will be staffed with three personnel (at minimum). The ladder will be staffed with four personnel (at minimum). The ambulance will be staffed with two personnel.

    (c) Search and rescue operations must be anticipated and will be of primary importance.

2. Operations:

    (a) The engine will advance a line to attack the fire.

    (b) The ladder company will initiate a two-team offense that will simultaneously operate on both stories of a two-story dwelling. The officer and forcible entry person will make entry on the first-floor level, then search the escape paths and rooms and shut off the electricity. The driver and outside vent person will use ladders to make entry into the second-floor bedroom. Following the search, the members will check for extension.

    (c) If additional ladder personnel are available, the fifth firefighter shall circle the house, checking in and around shrubbery for any occupants who may have been injured or overcome while escaping. This firefighter will also shut off the gas. When the search has been completed, he will assist in placing lights and fans. The sixth firefighter will bring lights, fans, and salvage covers to the front door.

Drill Outlines

Individual Responsibilities

1. Ladder company officer

    Tools: Radio, short hook, hand light, SCBA.

    Position: Entrance to the building.

    Duties: Makes a rapid size-up. Orders the members to initiate the SOP unless the fire conditions or collapse potential indicates otherwise. When outside entry operation to the second floor isn't required, he notifies the members to prevent breaking windows unnecessarily.

2. Forcible entry firefighter

    Tools: SCBA, ax, halligan tool, hand light.

    Position: Entrance to the building.

    Duties: Checks to see whether the door is locked. Forces entry into the building. Conducts a search with the officer and ventilates as directed by the officer. Shuts off the electrical utilities.

3. Outside vent firefighter

    Tools: SCBA, radio, short hook, hand light.

    Position: On the floor above the fire.

    Duties: If possible, advances to the second floor by the interior stairs. Otherwise, if there is a porch- or garage-roof access to an upper-floor window, raises a ladder to this roof. Removes the entire window glass and frame, enters, and begins the search. Advises the officer by radio when he makes entry and when he leaves. On a one-story building, stands by to vent from the outside when instructed by the officer.

4. Driver

    Tools: Pickhead ax, portable ladder.

    Duties: Assists the outside vent firefighter in placing the ladder. Butts the ladder and remains in sight of the window to assist the vent person if necessary. If a victim is to be removed over the ladder, the driver calls for assistance and climbs the ladder to bring down the victim. If the outside vent firefighter enters through the interior stairs, the driver places a ladder to the sill of a window chosen by the outside vent firefighter.

5. Fifth firefighter

    Tools: SCBA, hand light, six-foot hook, gas shutoff tool.

    Duties: Searches the exterior of the building. Looks for signs of victims inside windows. Looks in shrubbery for any victims that may have collapsed during escape. Shuts off the gas at the exterior meter. Reports to the incident commander on the fire conditions and exposures 2, 3, and 4. Assists in moving fans and lights into the building.

6. Sixth firefighter

    Tools: SCBA.

    Duties: Brings two lights, two electric fans, a power cord, two cable reels, and two salvage covers to the entrance. Assists in placing the covers, lights, and fans in the building.

Search and Rescue

If, as a ladder team member, you meet with fire at the door, do not wait to advance behind the engine crew. Make access into rooms not involved by fire. Such access shall be made by complete removal of the window glass and frames. At night, place emphasis on the bedrooms.

Remove the top-floor window first to reduce the potential for flashover or backdraft. Stand to one side of the window when venting. Have your SCBA on. Position yourself upwind and do not get in a position where fire blowing out of a window would prevent your escape. Advise Command by radio that you are entering. Advise him of the quadrant that you are entering.

On entering the room, if heavy smoke and heat are pushing in, shut the door to allow you to make the search.

If you find a victim, the first consideration for removal should be by the interior stairs. Use the portable radio to get help. Remember which quadrant you are in.

Report the completion of the search and your exit to Command.

Venting and Opening Up

To check whether a structure is balloon frame, remove the baseboard from an exterior wall in an area away from windows and doors. If there is no sole plate present, assume balloon-frame construction.

Debriefing

The riding assignment board will be modified to reflect the job that each member must do. Make sure you know how to do your assigned job.

Consider putting labels at each riding position listing the duties and tools to be carried by the member in that seat.

Notes

What went right:

_____
_____
_____
_____

What went wrong:

_____
_____
_____
_____

What to do differently next time:

_____
_____
_____
_____

Volunteer Training Drills

Figure 23-1

Drill Announcement Flyer

COMING SOON TO YOUR NEIGHBORHOOD

On _____ at about _____ ,

we will be conducting a drill at

_____

IT IS ONLY A DRILL

The purpose of this drill is to practice the skills that we will use in the event of an emergency in your neighborhood. The drill will last about _____ hour(s).
We will be placing hoselines in the street for a short period of time.

If this drill causes any hardship or inconvenience for you, or if you have any special concerns, contact _____ at _____
before _____ .
We will be happy to work with you to keep any inconvenience to a minimum.

 REMINDER: Did you remember to change your smoke detector battery when you changed to and from daylight savings time?

24. Advancing Hoselines by Ladder

Objective

To develop proficiency in advancing hoselines to the upper floors of a structure and extending lines downstairs to extinguish a simulated basement fire.

Setup Time

This drill is best accomplished at your jurisdiction's fire training academy as part of a live-fire training exercise. It requires minimal setup time. However, the extension of hoselines, without live fire, can also be accomplished on a local building, with the owner's permission, or on a multistory fire station.

Materials Required

Engine and truck company.

Rescue dummy.

References

"Methods for Safe Aerial Device Operations," John Mittendorf, *Fire Engineering*, June 1996, page 79.

Preparation

Place the rescue dummy in the building for Evolution No. 3.

Running the Drill

Evolution 1

A four-person team pulls an attack line from the engine and clears the line from the bed. They advance the line up exterior stairs to the second floor. After making entry, they advance the line downstairs and knock down the fire. They then back the line out of the building.

Evolution 2

The team raises a ground ladder to a wide second-floor window and opens it. They then reset the ladder for entry through the window and advance a line through the window to the second floor. They advance the line downstairs and knock down the fire, then back the line out of the building.

Evolution 3

The team raises a ground ladder to a second-floor window that has a narrow opening, positioning the ladder for rescue. They advance the line through the window, then downstairs to knock down the fire. They then back the line out of the building and place the rescue dummy on the ladder for rescue.

Evolution 4

The members raise an aerial ladder to the roof and open a roof vent. They position a ground ladder to enter through the roof, then advance an attack line through the roof vent to extinguish the fire.

Predrill Briefing

"The purpose of this drill is to develop proficiency in getting attack lines into a building via ladders. There will be four evolutions, which you will perform wearing full gear and breathing apparatus."

Debriefing

The debriefing is based on observations made using the Advancing Lines Checklist (Figure 24-1).

Notes

What went right:

_____
_____
_____
_____

What went wrong:

_____
_____
_____
_____

What to do differently next time:

_____
_____
_____
_____

Figure 24-1

> # Advancing Hoselines Checklist
>
> **Objective 1—Personal Safety Checks**
>
> > Check SCBA tank pressure against regulator pressure.
> >
> > Check SCBA low-pressure alarm.
> >
> > Check SCBA mask seal and exhaust valve.
> >
> > Turn on PASS device.
> >
> > Make sure gloves are on and ear flaps are down.
>
> **Objective 2—Advancing Lines**
>
> > Pulls line smoothly.
> >
> > Clears line from the bed.
> >
> > Advances line smoothly.
> >
> > Properly drapes line across shoulder while climbing ladder.
> >
> > In stairwells, the team flakes line upstairs past doorway before advancing.
> >
> > The team clears kinks in the hoseline.
>
> **Objective 3—Fire Attack**
>
> > Bleeds air from hose.
> >
> > Selects correct stream pattern.
> >
> > Attack team stays low.
> >
> > Proper position on the lines (protects victim).
> >
> > Makes quick hit on ceiling to kill heat.
> >
> > Keeps stream moving.
> >
> > Avoids walls to minimize steam.
> >
> > Backs out of fire room, facing the fire.

25. Drowning Response

Objective

To learn what actions to take at the scene of a reported drowning, both prior to the arrival of the dive team and in support of the dive team. The members walk through a dry run.

Setup Time

Preparatory meeting with the dive team coordinator.

Advance notice to media and residents near the drill site.

Obtain keys for any gates that control access to the drill site.

Materials Required

Blind caps for hose. One cap should have an air fitting so that the hose can be filled with compressed air.

Personal flotation devices (life jackets).

Throw ropes.

References

Local SOP for water rescue.

Preparation

1. Meet with the dive team representative to explain the purpose of the drill. Go over this outline and adapt it to any local protocols.
2. Prepare and distribute notices of the drill to the media and residents near the drill location.

Running the Drill

The drill takes place in two parts: first, a lecture describing the actions to be taken by the first-arriving units; and second, a dry-run response to a simulated drowning incident at a local pond or lake site.

Predrill Briefing

"The purpose of this drill is to develop and walk through the response to a drowning incident. The basic principles will be the same for an ice rescue scenario. The procedures will be described by the dive team representative."

Lecture Topics and Key Points

Information to Be Gathered at the Scene

1. Separate witnesses from other bystanders and from each other.
2. Obtain from witnesses:
   (a) The last location that the victim was sighted.
   (b) The number, age, and gender of the victims.
   (c) The time last sighted.
   (d) A description of the victim's clothing.
   (e) The place that the victim entered the water.
3. Techniques to establish the last-sighted position:
   (a) Set up range marks. Have the witnesses return to the place they were standing when they last saw the victim. Have them identify a feature on the opposite shore in line with the victim's last position. Place a cone or other easily visible marker at the witnesses' position and one at the reference position.
   (b) Estimate the last position from the reference points nearby. (Example: "20 feet south of the center of the bridge," or "halfway between the rock and the shore.")

Describe the Standard Dispatch for a Drowning Incident

1. Governed by (your local) SOP No. _____.
2. List the units to be dispatched.
   (a) Optional: Do you need to request a medevac helicopter and set up a landing site?
   (b) Optional: Do you need to request a command unit, a school bus, or a squad for diver rehab in cold weather?
3. Describe the expected response time for the dive team.

Review the Initial Actions to Be Taken and Not to Be Taken

1. There will be tremendous pressure on the members to "do something." DO NOT enter the water without a personal flotation device and water rescue training. Don't do anything that you haven't been trained to do.
2. If the victim is visible:
   (a) Attempt to throw a line to the victim.
   (b) Use the blind caps to inflate hose sections, and float them out to the victim.
   (c) Remove any road barriers and clear the way for the first-arriving boat.
   (d) In cold weather, try to keep the victim talking to maintain mental alertness.

3. If the victim is not visible:
   (a) Separate the witnesses and obtain information as previously described.
   (b) Ask bystanders where they were standing when they last saw the victim. Then, establish marks on the shore opposite the last-sighted position.
   (c) Request the public affairs officer and CISD team to assist with the public.
   (d) Choose the boat launching site that is closest to the victim's position. Rope off the area. Place one or two salvage covers on the ground as dive equipment staging areas.
   (e) Clear parking areas for the following vehicles (per your SOP): engine company, ambulance, medic unit, boat trailer, command unit, dive team personal cars, and dive team truck and boat.
   (f) Remove any barriers on the access road to the boat launch site.
   (g) Establish a landing site for the medevac helicopter.
   (h) Request an extra ambulance to assist the victim's relatives and friends.
   (i) Assign an officer or senior firefighter to act as liaison to the victim's friends and relatives if the PIO is not on the scene. This officer should reassure those people that extra help is coming.
   (j) Assign personnel to support the dive team. Anyone working at the water's edge shall wear a PFD. Also, assign a member qualified to operate a cascade to stand by to fill bottles.

Debriefing

This is a response for one location. What other locations are likely drowning sites, and what might you have to do differently there?

Notes

What went right:

_____
_____
_____
_____

What went wrong:

_____
_____
_____
_____

What to do differently next time:

_____
_____
_____
_____

26. Swimming Accidents

Objectives

To practice the techniques for removing a diving accident victim from a swimming pool.

Setup

Coordinate the logistics of the drill with your local pool manager.

Coordinate your response with the response of the pool's lifeguards.

Materials Required

Backboard.

Immobilization materials.

A change of clothes for members entering the pool. Warn them ahead of time to wear old shoes that can be worn in the pool.

Arrange for a pool lifeguard to act as a victim.

References

Essentials of Emergency Care, David Limmer, Bob Elling, Michael O'Keefe, and Edward T. Dickinson, M.D., Prentice-Hall Inc., 1996, Chapter 18.

Preparation

1. This drill requires meeting with the pool management and lifeguard staff. Emphasize that the purpose of this drill is to ensure that everyone understands what to do and that they work together to care for the victim. It is not a race between the lifeguards and the fire department.

2. Arrange with the pool manager to jointly describe to any spectators what is going on. An outline of the points that you should make as the fire department spokesperson is included below.

Running the Drill

1. Stage the responding ambulance and engine company a short distance from the pool location. Station an officer with a portable radio at the pool.

2. The pool manager announces to the patrons that there will be a diving accident drill and that the patrons shall clear the pool area when the drill starts.

3. A lifeguard dives into the pool and simulates a victim. The pool's lifeguards respond. After a 30-second delay, the officer has the units respond.

Predrill Briefing

"This drill is to practice your response to a diving accident. The pool management is participating in this drill and will stage a diving accident. The pool lifeguards will have already reached the victim by the time you arrive. You will assist in getting the victim secured and fully immobilized on a backboard. You will conduct a survey and prepare for transport. You should be prepared to enter the water, dressed as you would be during an actual response. It would be a good idea to change into clothes and shoes that you don't mind getting wet and to leave your wallets in your lockers or with the driver of your unit.

"This drill will take place while the pool is open for business. If you need to go into the water to assist, remember to enter the pool gently. The simulated victim will be a pool lifeguard. In the interest of good relations, secure the victim to the board with roller bandage. Do not use tape for this exercise."

Presentation Outline

"We would like to thank the management of _____ for allowing us to conduct this joint drill. A diving accident is a very serious thing. Injury to a spinal cord can change a person's life instantly. Such victims may no longer be able to walk, use their arms, or control their bodily functions. Only infrequently can spinal cord injuries be repaired. Some surgeons say that the worst thing that happens to some victims is that they live. That may sound very harsh, but that is the way it is.

"The purpose of today's drill is to practice our handling of a diving accident so that we can give a victim the best chance of recovery. The first actions taken by the lifeguards are very important in preventing further damage. What we are doing now is getting the person secured to a backboard that will hold the spine in position while the victim is transported to the hospital.

"The rescuers will be paying attention to keeping the spine straight and securing the person's head so that no more damage will be done if the person tries to look around. In a real incident, we would use tape to secure the person's head. In this drill, we are using roller bandage because our victim would prefer to go through the rest of the summer with his eyebrows intact.

"Once the victim is stable on the board, we would supply him or her with oxygen and start an IV as ordered by the doctors. Our paramedics have direct communications with the hospitals and can start medications right here.

"At the same time that this is going on, the engine company will be setting up a landing site for the medevac helicopter.

Volunteer Training Drills

"Everyone likes to swim and dive, and we do not want to spoil anyone's fun. We do need your help, however, in preventing these types of accidents. We ask that you never dive into water that is not specifically set up for diving. If you are going to leap into a pond or lake or even a pool, always make the first jump feet-first. That way, the worst that can happen is a broken ankle, not a broken neck.

"We would like to thank all of you for allowing us to come here. Our primary purpose in being here today is to practice so that we will be better able to help in the event of an accident."

Debriefing

What went right:

_____
_____
_____
_____

What went wrong:

_____
_____
_____
_____

What to do differently next time:

_____
_____
_____
_____

27. Area Familiarization

Objectives
To review response routes.
To ensure knowledge of street locations in the first-due area.
To ensure knowledge of water-source locations.

Setup Time
About one hour using company map books.

Materials Required
Company map books.
Correction fluid.
Transparency film.
Transparency markers.
Access to a copy machine.
Overhead projector.
Pencils for the members.

Preparation
1. Make a copy of the company map book pages for the streets within one mile of the station. Also, make copies of the preplan drawings for the major public buildings in the first-due area.
2. Use correction fluid to remove street and building names and any special water source symbols, such as dry hydrants, from the street map copies. Remove the symbols for standpipes, sprinklers, and alarm panels from the preplan building drawings.
3. Make a copy of the blank pages for each member, and also make a transparency to use during the drill.

Running the Drill
1. Provide the members with the maps and pencils. Give them about 15 minutes to fill in the missing information.
2. Project your transparency onto the screen, and use your markers to fill in the missing information. Let the members tell you the missing street names and symbols.

Predrill Briefing
"There are any number of opportunities for error when you are responding to an alarm. You may get lost or not go to the right part of a building. The purpose of this drill is to review that information for the streets within one mile of the station and for major buildings in the first-due area. In the process, you will get the chance to start building your own personal map book of the first-due area."

Volunteer Training Drills

Notes

What went right:

_____
_____
_____
_____

What went wrong:

_____
_____
_____
_____

What to do differently next time:

_____
_____
_____
_____

28. Apparatus and Equipment

Objectives
To review the location of all tools on the apparatus, as well as how to inspect and operate them.

Setup Time
About one hour to make a drawing and copies for each member.

Materials Required
An outline drawing of each station apparatus, showing the location of each compartment.

Transparency film.

Transparency markers.

Access to a copy machine.

Overhead projector.

Pencils for the members.

References
"How Basic Can You Get? How About Hand Tools for Truck Work?", Tom Brennan, *Fire Engineering*, June 1994, page 152.

"Forcible Entry Tool Storage and Maintenance," Ray McCormack, *Fire Engineering*, January 1994, page 28.

Preparation
Make a copy of the outline drawings for each member. Have a pencil for each member, plus a few extras.

Running the Drill
1. Provide the members with the drawings and pencils. Give them about 15 minutes to list the equipment carried in each compartment.
2. Project your transparency onto the screen, and use your markers to fill in the missing information. Let the members tell you what is in each compartment.
3. Pull all of the equipment and inspect it. Operate all of the power equipment.

Predrill Briefing
"We carry a lot of different equipment, some of which doesn't get used very often. The purpose of this drill is to ensure that everyone knows what we carry and where to find it. We will then take the equipment out, inspect it for damage and proper operation, and make sure that every member knows how to fuel and operate all of the power tools. Please remember that different power equipment requires different oil/fuel ratios. If you use the wrong mixture, you will damage some very expensive equipment."

Volunteer Training Drills

Debriefing

Some power equipment requires only a couple of steps to get started. Others require several steps. Discuss painting a number beside each part that must be adjusted to start the tool. The highest number goes on the starter cord. If the starter cord is numbered 5 and the member hasn't done four other things, pulling on the cord probably won't get that particular tool to start.

Use the debriefing period and the reference articles to lead a discussion of equipment storage. Is there a more efficient way to store equipment?

Notes

What went right:

_____
_____
_____
_____

What went wrong:

_____
_____
_____
_____

What to do differently next time:

_____
_____
_____
_____

Figure 28-1

Tool Inspection Checklist

Hand Tools

    Clean tools to uncover hidden problems such as cracks or burns.

    Heads on cutting and striking tools are tight on handle.

    Ax heads are free of paint and are lightly oiled.

    Ax heads are sharpened to a proper edge—not too sharp, not too dull.

    Edges of striking and cutting tools are free of burrs.

    Handles are free of cracks.

    Tool ID marking (paint or tape) is in good condition.

Power Tools

    Inspect for damage to covers, handles, guards.

    Blades are sharp and free of damage.

    Drive belts have correct tension.

    Power tool is full of fuel.

    There is a spare fuel can for each gas/oil mix. Each can is marked and full.

    Power saws return to idle and the blade stops moving when the trigger is released.

    Hydraulic tools are stored so as to protect the hoses.

    Hydraulic hoses are free of cuts and kinks.

    Examine cutting torch for any damage to hoses, controls, gauges, or cylinders.

Volunteer Training Drills

29. Ventilation

Objective

To review tool operations for ventilation.

Setup

Try to obtain access to a structure scheduled for demolition. If this isn't possible, an afternoon will be required to locate a supply of wood pallets, plus an hour to review your power tool manuals. The setup before the drill requires about 30 minutes.

Materials Required

Wood pallets.

Power saws and operating manuals.

Axes and sledgehammers.

Fans.

References

"Attic Ventilators Affect Firefighter Safety," Harry J. Oster, *Fire Engineering*, May 1998, page 107.

"Strip Ventilation Tactics," John Mittendorf, *Fire Engineering*, March 1996, page 49.

"Ventilation Problems," Bob Pressler, *Fire Engineering*, April 1995, page 102.

"Safety During Roof Operations," Gene Carlson, *Fire Engineering*, November 1994, page 12.

"How Basic Can You Get? How About Hand Tools for Truck Work?", Tom Brennan, *Fire Engineering*, June 1994, page 152.

"Venting," Tom Brennan, *Fire Engineering*, March 1994, page 178.

Fire Officer's Handbook of Tactics, Second Edition, John Norman, Fire Engineering, 1998.

Preparation

1. Procure a supply of wood pallets that are old but in reasonable condition.
2. Select a site near a dumpster where you can dispose of the pallets once they have been cut up. Make sure that you have permission to use the dumpster.

Running the Drill

1. Conduct a review of saw operations.
2. Inspect the axes and sledgehammers. Use the checklist from Drill 28, Apparatus and Equipment.
3. Conduct a practical session using both axes and power saws to cut up the pallets.

Predrill Briefing

"This drill involves practice with venting tools. You will inspect the tools, review the use of power saws, and get some hands-on practice. It is incredibly easy to lose balance or to walk off the edge of a roof. Practice using a backup person—someone who is watching out for you while you make your cuts."

Outline for Ventilation Drill

Power Saw—Circular Blades

Two-cycle engines require the proper mixture of oil in the fuel. Not all saws use the same mixture.

Check the belt tension. The belt should deflect its own thickness.

Look for cracks in the belt.

Look for cracks in the saw handles.

The handles must be dry, clean, and free of oil.

Check that the power arm screws are tight.

Check the power arm for cracks.

Inspect the cutoff wheels.

(a) Look for any cracks.

(b) Look for any delamination.

(c) Check for paper "washer" (label) around the mounting hole. This washer distributes the load evenly to the wheel.

(d) Conduct a ring test. A flat sound means a damaged wheel.

(e) Do not store wheels in the same compartment as gasoline. Doing so will cause a wheel to deteriorate.

Check the fit of the cutting wheel. It should be a loose fit. Never force a cutting wheel onto the spindle. The wheel should be tight enough so that it doesn't slip but not overtight. Overtightening warps the flange and changes the contact area. The wheel should turn freely and must not touch the guard.

When installing a new wheel, run it at full speed for 30 seconds before starting the cut.

Carbide-Tip Blades

(a) Do not use if one or more tips are missing.

(b) The blades can be repaired—don't throw them away.

(c) Carbide blades are more likely to throw debris—up to 100 feet.

(d) Carbide blades are more likely to kick back.

Correct Stance

Demonstrate the position of the guide person and the correct signals to use.

(a) 1 = stop; 3 = back up.

(b) Always back up—never cut forward.

Demonstrate guide adjustment. Talk about kickback.

Starting

(a) Stop switch.

(b) Choke switch.

(c) Throttle.

(d) Prepull the cord to stop, then pull sharply.

(e) Check the idle speed. The blade should not be turning at idle.

(f) Rough running and smoky exhaust most likely mean that the air cleaner element is dirty. Check that the air passages are clear.

(g) Carburetor is a French word meaning "Leave it alone!"

Safety Tips

(a) Do not climb with the saw running. If the engine races, the blade will start turning.

(b) Use FDNY practice to move a running saw around—turn it like a wheelbarrow.

Operation

(a) Adjust the guard so that the rear section is close to the work piece.

(b) Cutting with a wheel: Bring the wheel into contact with the engine running at slow speed. Increase the speed as the wheel cuts into the material. Maintain pressure on the wheel to get the best cutting action. Do not twist the wheel in the cut.

(c) Cutting with a carbide blade: Run the saw up to full speed before contacting the surface. Let the saw do the cutting—do not force the blade into the material.

Chain-Type Saws

Check the operator's manual for your particular saw.

Generally, the chain shouldn't have any slack when the saw is cold.

Like the carbide blade, the chain saw should be running at high speed when it contacts the material.

Be especially careful not to let the tip of the chain saw touch anything. Emphasize the dangers of kickback.

Review the sequence of cuts for making a roof opening.

Practice on the pallets.

Volunteer Training Drills

Notes

What went right:

_____
_____
_____
_____

What went wrong:

_____
_____
_____
_____

What to do differently next time:

_____
_____
_____
_____

30. High-Rise Response

Objectives

To develop and practice the procedures for attacking a fire in a high-rise.

To establish the staffing required to carry out an attack on the top floor, using the stairs for access.

To develop a time line to stage an attack team and a backup team to extinguish the fire.

To practice the procedures for charging the standpipe system using a lower-floor discharge connection.

To practice the procedures for maintaining an air supply for the attack teams.

To develop the logistics to protect hoselines and personnel from falling glass.

Setup

This will take several meetings with the management of a local high-rise building and company officers to plan the drill. Once planned, no additional setup time will be required.

Materials Required

Preplans for the high-rise building.

Your local SOP for implementing the ICS.

Air cascade for recharging bottles.

Press releases explaining the purpose of the drill.

Three watches and clipboards.

A tape recorder or other means of timing and recording communications.

In terms of personnel, this drill also requires five training officers, a public affairs officer, and fill-in companies.

Volunteer Training Drills

References

"Firefighting Operations in High-Rises Under Construction," Jerry Tracy, *Fire Engineering*, December 1997, page 47.

"Information—The Key to Successful Firefighting," Vincent Dunn, *Fire Engineering*, October 1996, page 32.

"The Standpipe Stretch Mindset," Ray McCormack, *Fire Engineering*, October 1996, page 64.

"High-Rise Firefighting Strategies," Vincent Dunn, *Fire Engineering*, August 1996, page 32.

"High-Rise Fire," Bob Pressler, *Fire Engineering*, April 1996, page 28; August 1996, page 124.

"Checking for Fire Spread at a High-Rise Fire," Vincent Dunn, *Fire Engineering*, March 1996, page 26.

"Advanced Pumping Skills: Tandem and Dual Pumping," Leigh Hollins, *Fire Engineering*, February 1996, page 47.

"Testing Operational Procedures: Cooperation is the Key for Full-Scale Simulations," Bernard Dyer, *Fire Engineering*, January 1996, page 22.

"Danger of High-Rise Firefighting," Vincent Dunn, *Fire Engineering*, January 1996, page 30.

"Banker's Trust Fire, New York City," Ellsworth Hughes, *Fire Engineering*, June 1994, page 86.

"High-Rise Operations: Surviving Above the Fire," Gerald Grover, *Fire Engineering*, March 1993, page 22.

Fire Officer's Handbook of Tactics, Second Edition, John Norman, Fire Engineering, 1998.

Preparation

1. Meet with the management of the largest high-rise building in your response area. Request access to the building on a weekend or evening to conduct the drill. Explain that you will primarily be operating in the stairwells and won't cause any water damage to the building. Request permission for firefighters to conduct a simulated search in the hallways of the top floor with the lights turned off. Your purpose is to set up and practice a procedure to protect the building and occupants in case of an emergency. The specific emergency is a basement fire that extends up an elevator shaft and fills the building with smoke.

2. This will require at least a two-alarm assignment. Make arrangements with mutual-aid companies to cover those units that will be committed to the drill.

3. This drill will be physically challenging to the participants. Make sure that there is EMS support at the scene and that you have access to an elevator in case of a medical emergency.

4. Read through the drill as outlined below. Assign training officers to monitor and record times for each of the seven items listed.

5. If possible, place a tape recorder with the training officer at the command post. That officer should announce times at two-minute intervals so that the flow of communications can be reviewed.

Running the Drill

1. Have the companies respond to the building from a nearby staging area and establish the incident command system.

2. Advise the incident commander that a fire in the basement is being controlled by an operating sprinkler system but that it has spread to the top floor through a cableway, and people are reported trapped on the top floor. The elevators are out of service. Breathing apparatus will be required from the middle floor to the top floor. The fire department connection is unusable. The connection will have to be made inside the building. Record the time.

3. Have a training officer stationed at the midlevel landing (for example, the sixth floor of a 12-story building). This TO should instruct the attack team to don SCBA for the rest of the climb. (If your first drill is in a 40-story building, you may not want to have the teams climb 20 floors in SCBA!)

4. Have the training officer at the top floor advise Command when the team reaches the top. Record the time at the command post. Record the air bottle pressures. The team is then instructed to enter the hallway and conduct a search (crawling) for 15 minutes. They are not to force doors or damage the building. They are to retreat when their low-pressure alarms sound.

5. Toward the end of the 15-minute period, the training officer on the top floor radios a Mayday for the team trapped with low air. The time of the Mayday is recorded at the command post. The training officer on the top floor advises Command when a rescue team reaches the top floor and initiates a search. The time is recorded at the command post.

The challenge of this drill is to accomplish the following:

1. To place an attack team and a backup team on the top-floor landing of the stairwell and keep them there for 15 minutes in full gear and SCBA.

2. When faced with an unusable fire department connection for the sprinklers and standpipe, to establish a connection using a lower-floor discharge connection.

3. When faced with the possibility of falling glass, to establish a means of protecting hoselines and personnel.

4. To test the ICS operation, maintaining accountability of personnel and knowledge of their location.

Volunteer Training Drills

Determine:

1. How long does it take to get the attack teams in place?
2. How much air do the attack teams have left when they get there?
3. How do you set up an air bottle relay to resupply the attack teams? How many personnel does it take, and how long does it take to get it set up?
4. How much air is required for the attack teams to operate for 15 minutes on the top floor?
5. When faced with the problem of falling glass, where do you get plywood to protect the hoselines? How long does it take to get it there? Where do you position the apparatus for protection from falling glass? How do you move people in and out of the building?
6. How well does the ICS work? Were the right sectors established? Was accountability maintained?
7. Given a Mayday call from the top-floor team, who would respond? Was there a ready assist team identified? Where were they staged? How much air did they have left when they arrived at the top floor? How did they get their air supply replenished?

Predrill Briefing

"The purpose of this drill is to test the procedures for operating in a high-rise under severe smoke conditions. Two teams on the top floor will conduct a search. The key problem will be your air supply.

"When you enter the building, you will have to use the stairs to reach the top floor. A training officer will tell you when to start breathing from your bottles. When you reach the top floor, another training officer will record your bottle pressures. You will then be asked to conduct a search of the top-floor hallways, crawling along the floors and taking a hoseline with you. The hoseline will not be charged, and you must be careful not to damage the building. Stick to the hallways and do not try to enter any offices or to force any doors. Keep moving so that we can test how well the department is able to establish and maintain a system to provide fresh bottles to the attack teams. When your low-pressure alarms sound, return to the stairway."

Debriefing

Use those items that you were measuring and observing as the basis for the debriefing. Get input from the attack crews and the assist crews. Get input from the training officers on the top floor. What shape were the crews in after their climb?

Discuss in detail the flow of air bottles to the teams. How many teams have to be rotated from halfway up the building to the top floor to keep the search going for 15 minutes?

What could you do to improve the situation?

If you had to vent to accomplish the task, how would you do it? Who gives the approval and the order to vent?

What kinds of problems were experienced in keeping track of people and logistics? Was somebody clearly in charge of these functions?

Find the problems, propose solutions, and rerun the drill in a different building.

Notes

What went right:

_____
_____
_____
_____

What went wrong:

_____
_____
_____
_____

What to do differently next time:

_____
_____
_____
_____

31. Multiple-Casualty Incidents

Objectives

To practice multiple casualty rescue operations.

To develop SOPs for multiple-casualty operations.

To practice triage operations.

To practice using the ICS.

Setup Time

There are several scenarios: a school bus accident; a bleacher collapse at a Little League softball field; and a collapse of merchandise racks at your local supermarket, furniture store, or warehouse-type store.

The easiest one to set up is the Little League bleacher collapse. The second easiest is the school bus—if you can borrow one from the local school board.

You will need to meet with a service organization or some relatives of members to obtain volunteer victims.

Materials Required

Access to a site.

Victims.

Triage tags and protocols.

Injury makeup and labels or a moulage kit.

Press releases.

Release forms for all of the victims.

References

The local SOPs for multiple-casualty incidents.

"California Highway Disaster," Mark H. Meaker, *Fire Engineering*, May 1998, page 51.

NFPA 1470, *Search and Rescue Training for Structural Collapse Incidents*.

Drill Outlines

Preparation

1. This is an excellent drill to involve a local service group, such as the Girl Scouts or Boy Scouts. Meet with that group and explain that you will need 12 to 15 individuals to serve as victims. Each victim will have makeup applied to simulate injuries and will be instructed on what to say to the rescuers. Each victim shall understand that the rescuers will do a full-body survey, but they will be instructed not to remove any clothing. If the victims can wear old clothes on which the sleeves and pant legs can be cut, they should do so. The parents of Boy and Girl Scouts should be available to participate as victims and to monitor the drill activity.

2. The injuries can be simulated with standard makeup. A Boy Scout supply store can provide black, blue, white, and red theatrical makeup. Small pieces of wood taped to the arms and legs or a small piece of wood taped to the spine will give realistic indications of problems that should be picked up in the survey. Victims with simulated broken bones should be advised to scream appropriately if those body parts are moved. At least one victim, with makeup applied to simulate a pale face, should be placed sitting down a short distance from the major group of victims. This victim should be instructed to act in a disoriented manner (doesn't know who he is or what happened) when and if he is located and identifed as a victim.

3. Try to coordinate this drill with your local medical facility so their personnel can drill their disaster plan along with your incident.

Running the Drill

1. Have the victims report 90 minutes before the scheduled start for instructions and makeup. Be absolutely sure that everyone participating signs a release form. If children are involved, ensure that the form is signed by their parents or guardians. If possible, have that person on site to monitor the child. The parents can also be very effective, panicky participants: "I was in the rest room, and when I came back, the stands had collapsed!" Mark with duct tape those victims wearing clothes that can be cut.

2. Responding apparatus should be staged at a nearby location and dispatched by a training officer so that they don't all arrive together. Resources should be staged to provide an appropriate number of personnel and supplies to manage the incident.

Volunteer Training Drills

3. Observe the operation of the command post and keep a timed record noting:

    (a) The steps taken to ensure the stability of the collapse site.

    (b) When the first Priority One patient was identified.

    (c) When the last Priority One patient was identified.

    (d) The proper use of triage procedures.

    (e) The proper assignment of treatment priorities.

    (f) When the Priority One patients were transported.

    (g) When the disoriented victim in shock was located and treated.

    (h) The assignment of sector officers and a safety officer.

    (i) The time to treat and transport all of the victims.

    (j) The proper tracking of transportation—i.e., who went where, and what types of injuries went where.

Predrill Briefing

"The purpose of this drill is to practice multiple-casualty treatment procedures. Apparatus will be staged a short distance from a simulated disaster and will be dispatched by a training officer. Your task will be to implement the ICS, establish order, and triage and treat the victims. Use triage tags, bandage and splint the wounds, and immobilize patients as necessary.

"This drill uses live victims who have volunteered to participate. Be especially careful of these volunteers. They have simulated injuries. You do not want them to leave with real injuries! You are not to remove the victims' clothing. Victims whose pant legs or shirt sleeves can be cut have a ring of duct tape on the sleeve or cuff. Respect the dignity and modesty of the victims. Both parents and the press will be monitoring your actions.

"If a victim's injuries call for application of oxygen, place the mask loosely around the victim's neck. Do not flow oxygen. If the injuries call for an IV, set up a bag and line, then secure the line to the arm with tape. Do not start any IVs."

Debriefing

The debriefing should cover the 10 evaluations listed above. Give the victims and their parents an opportunity to tell you their observations. It is important to know how they perceive your actions. If their perceptions are incorrect, you will have a chance to explain your actions and to build good public relations. You'll also have a chance to think about ways to prevent others from gaining those same perceptions in future incidents.

Notes

What went right:
_____
_____
_____
_____

What went wrong:
_____
_____
_____
_____

What to do differently next time:
_____
_____
_____
_____

32. Haz-Mat Operations With Multiple Casualties

Objectives

To practice SOPs for haz-mat incidents.

To practice multiple-casualty rescue operations.

To practice SOPs for multiple-casualty operations.

To practice triage operations.

To practice ICS operations.

Setup Time

You will need to locate a building with a loading dock where you can place a few empty 55-gallon drums and stage a few victims.

You will need to meet with a service group or the relatives of members to obtain some volunteer victims.

Materials Required

Access to a loading dock or simulated loading dock.

Victims.

Triage tags and protocols.

Injury makeup and labels or a moulage kit.

Press releases.

Release forms for all of the victims.

A few empty drums with appropriate haz-mat markings.

Baking soda and vinegar.

Overpacking equipment for the haz-mat drums.

Decontamination equipment.

References

"Haz Mat: On the Line," Stephen L. Hermann, *Fire Engineering*, April 1998, page 103.

"Burning Gasoline Tankers: The Best Action May Be No Action," Peter M. Stuebe, *Fire Engineering*, November 1997, page 41.

"Dry Cleaners: Miniature Chemical Plants," Dan Pontecorvo, *Fire Engineering*, May 1997, page 73.

"Haz-Mat Response to Leaking Battery Acid," Peter M. Stuebe, *Fire Engineering*, May 1997, page 50.

"Testing Operational Procedures: Cooperation is the Key for Full-Scale Simulations," Bernard Dyer, *Fire Engineering*, January 1996, page 22.

"Evaluating the Effectiveness of Haz-Mat Decontamination," David Peterson, *Fire Engineering*, April 1994, page 75.

"The Unique Challenge of Urban Haz-Mat Fires," Denis Onieal, *Fire Engineering*, February 1993, page 27.

"Haz-Mat Notebook," Larry Radtke, *Fire Engineering*, February 1993, page 10.

"Swimming Pool Disinfectants: Seasonal Usage and Storage Hazards," Patrick Clarke, David Szymanski, and John Sachen, *Fire Engineering*, July 1992, page 73.

"Derailment at Seacliff," Mike Proett, *Fire Engineering*, April 1992, page 56.

"Using Community Right-to-Know Data," Steve Faelz, *Fire Engineering*, February 1992, page 75.

"Phosphorous Incident in Cooper County, Missouri," Bill Simmons, John Saechen, Leo Tierney, and John Coburn, *Fire Engineering*, July 1991, page 64.

"ICS Flow Chart," Larry Radtke, *Fire Engineering*, February 1991, page 10.

"Rule No. 1: Do No Harm," Stephen Hermann, *Fire Engineering*, February 1991, page 57.

NFPA 471, *Responding to Hazardous Materials Incidents*.

Preparation

1. This is an excellent drill to involve a local service group, such as the Girl Scouts or Boy Scouts. Meet with that group and explain that you will need 12 to 15 individuals to serve as victims. Each victim will have makeup applied to simulate injuries and will be instructed on what to say to the rescuers. Each victim shall understand that the rescuers will do a full-body survey, but they will be instructed not to remove any clothing. If the victims can wear old clothes on which the sleeves and pant legs can be cut, they should do so. The parents of Boy and Girl Scouts should be available to participate as victims and to monitor the drill activity.

2. The injuries can be simulated with standard makeup. A Boy Scout supply store can provide black, blue, white, and red theatrical makeup. Small pieces of wood taped to arms and legs or a small piece of wood taped to the spine will give realistic indications of problems that should be picked up in the survey. Victims with simulated broken bones should be advised to scream

appropriately if those body parts are moved. At least one victim, with make-up applied to simulate a pale face, should be placed sitting down a short distance form the major group of victims. This victim should be instructed to act in a disoriented manner (doesn't know who he is or what happened) when and if he is located and identifed as a victim.

3. Try to coordinate this drill with your local medical facility so that their personnel can drill their disaster plan along with your incident.

Running the Drill

1. Have the victims report 90 minutes before the scheduled start for instructions and makeup. Be absolutely sure that everyone participating signs a release form. If children are involved, ensure that the form is signed by their parents or guardians. If possible, have that person on site to monitor the child. The parents can also be very effective, panicky participants. Remember, if they are in the hot zone, they are also victims. Mark with duct tape those victims wearing clothes that can be cut.

2. Responding apparatus should be staged at a nearby location and dispatched by a training officer so that they don't all arrive together. Resources should be staged to provide an appropriate number of personnel and supplies to manage the incident.

3. Observe the operation of the command post, and keep a timed record noting:

   (a) The steps taken to establish hot, warm, and cold zones.

   (b) The steps taken by arriving personnel. If the steps are inappropriate, declare those personnel to be additional victims.

   (c) The location of apparatus in terms of distance, terrain, and wind.

   (d) When the first Priority One patient was identified.

   (e) When the last Priority One patient was identified.

   (f) The proper use of triage procedures.

   (g) The proper assignment of treament priorities.

   (h) When the Priority One patients were transported.

   (i) When the disoriented victim in shock was located and treated.

   (j) The assignment of sector officers and a safety officer.

   (k) The time to treat and transport all of the victims.

   (l) The proper tracking of transportation—i.e., who went where, and what types of injuries went where.

   (m) The proper containment of the hazardous material.

   (n) The proper decontamination procedures.

Predrill Briefing

"The purpose of this drill is to practice multiple-casualty treatment procedures when they are complicated by a haz-mat incident. The apparatus will be staged a short distance from a simulated disaster and will be dispatched by a training officer. Your task will be to implement the ICS, establish order, and triage and treat the victims. Use triage tags, bandage and splint wounds, and immobilize patients as necessary.

"Live victims have volunteered for this drill. Be especially careful of these volunteers. They have simulated injuries. You do not want them to leave with real injuries! You are not to remove the victims' clothing. Those victims whose pant legs or shirt sleeves can be cut have a ring of duct tape on the sleeve or cuff. Respect the dignity and modesty of the victims. Both parents and the press will be monitoring your actions.

"If a victim's injuries call for application of oxygen, place a mask loosely around the victim's neck. Do not flow oxygen. If the injuries call for an IV, set up a bag and line, then secure the line to the arm with tape. Do not start any IVs."

Debriefing

The debriefing should cover the evaluation criteria listed above. Give the victims and their parents an opportunity to tell you their observations. It is important to know how they perceive your actions. If the perceptions are incorrect, you have a chance to explain your actions and to build good public relations. You also have a chance to think about ways to prevent others from gaining those same perceptions in future incidents.

Volunteer Training Drills

Notes

What went right:
_____
_____
_____
_____

What went wrong:
_____
_____
_____
_____

What to do differently next time:
_____
_____
_____
_____

33. Railroad Emergencies

Objectives

To increase your awareness of the types of equipment and materials that move along your local rail lines.

To review the procedures for contacting railroad authorities.

To review the procedures for stopping trains during an emergency.

To review the procedures for accessing a trapped locomotive crew.

To review the procedures for using the emergency fuel shutoff and for setting the brakes on locomotives.

Setup Time

This drill requires contacting your local railroad safety offices and arranging a briefing. Be specific, and tell the representative that you want to cover at least the above points. Note that some components of locomotives maintain a charge of 2,500 or more volts after the engine and alternator have been shut down. Obtain the shutdown procedures for all of the models of locomotives that you are likely to encounter.

Materials Required

Screen and projector as required by the railroad representative.

References

"Haz-Mat Emergencies Involving Intermodal Containers," Gregory Noll, *Fire Engineering*, November 1997, page 63, and December 1997, page 61.

"Fatal Train Wreck in New Jersey," Martin McNulty and William Peters, *Fire Engineering*, May 1996, page 39.

"Safety at Freight Train Emergencies," Chuck Kenyon, *Fire Engineering*, May 1994, page 16.

Preparation

The preparation for this drill involves contacting the railroad's safety office, discussing the types of information you need for your members, arranging a time and place for the drill, and arranging for any visual aid equipment needed by your speaker.

Running the Drill

Running the drill consists of introducing the speaker and turning the drill period over to him. Send the railroad a thank-you letter after the drill.

Debriefing

Base your debriefing on your observations and the input of the railroad representative.

Volunteer Training Drills

Notes

What went right :

_____

_____

_____

_____

What went wrong:

_____

_____

_____

_____

What to do differently next time:

_____

_____

_____

_____

DRILL OUTLINES

34. Military Aircraft Emergencies

Objective

To learn special techniques for accessing injured crew members and the special hazards of military aircraft, particularly helicopters.

Setup Time

The easiest way to set up this program is to contact your local Air National Guard unit or other military air base. Request a briefing and an opportunity to inspect the aircraft. You can usually make arrangements to bring helicopters to a nearby athletic field. Be sure that you contact the right people and get permission to use the field.

Materials Required

Any materials that are requested by the military or that are required for your backup drill.

Preparation

1. Meet with the military representative, at least by telephone. Ensure that the representative knows that this is not a recruiting opportunity or a show-and-tell about the military. You should request specific information about accessing the crew compartment. You need to know:

   (a) How to approach the aircraft safely, out of line of any weapons discharge.

   (b) How to unlock and operate the access doors.

   (c) How to disconnect the batteries.

   (d) The location of any special sensors that may be heated or otherwise hazardous to your members.

   (e) The location of any hazardous materials, such as radioisotope power generators.

   (f) Any special recommendations on extinguishing agents.

   (g) How to activate any built-in suppression systems.

   Arrange the time and place, and find out whether the briefers need any special visual aid equipment.

Running the Drill

1. If helicopters are going to land at a local field, arrive at the field early. Survey the area for any debris that could be drawn into the engines or thrown about by the rotor wash. Place members around the field to prevent children from entering the landing area. Caution any civilians about the potential eye hazards of dust and debris during takeoffs and landings.

2. Running the drill involves introducing the military representatives. Since these drills are subject to the weather, have a backup drill ready to go.

3. Get the mailing address for the military command, and send a thank-you letter addressed to the commanding officer.

Volunteer Training Drills

Debriefing

Base your debriefing on your observations and the input of the military representative.

Notes:

What went right:

_____
_____
_____
_____

What went wrong:

_____
_____
_____
_____

What to do differently next time:

_____
_____
_____
_____

35. Hose Handling II

Objective

To develop proficiency in maneuvering hoselines.

Setup Time

15 minutes.

Materials Required

A flat, level parking area with a water source and good drainage.

Eight cinder blocks, painted white, to use as goal line markers.

A pumper, gated wye, four lengths of 1½- or 1¾-inch hose, and two matched nozzles.

55-gallon drum with a top plug (no side plugs).

Video camera.

A whistle for the training officer.

Preparation

Arrange a playing field as shown in Figure 35-1. Make sure that water can drain away without causing damage.

Running the Drill

The object of the drill is to develop teamwork in moving and directing a hoseline. The device is a 55-gallon drum, placed on its side, to be pushed across the goal line by a hose stream. Start the drum in the center of the field. If it goes out of bounds, restart it at the center field position. If the drum is too lively, put some water inside it to deaden its motion. If it's too slow, use a five-gallon can instead.

Use a whistle to stop the action. Have a safety officer, stationed at the gated wye, who can stop the water flow if the hose streams get out of control.

Predrill Briefing

"This is a drill to develop skill and teamwork in hose handling. It is a game of hockey using hose streams and a 55-gallon drum on a playing field. You will score a goal by keeping the drum in bounds and moving it across the goal line, which is marked by two cinder blocks.

"The hose streams are controlled by a safety officer at a gated wye. If you lose control of the hose streams, or use them on anything but the drum, the line will be shut down.

"A whistle means to shut down your hoseline."

Volunteer Training Drills

Debriefing

If you have a videotape record of the action, review that tape. How well did the teams keep the line straight behind the nozzle operator? Did the nozzleman have enough line out in front of him to control the nozzle, or did it get pulled back to his armpit? What happened when they tried to back up the line? Were the members spaced out well along the hoseline, or were they all bunched up and tripping over it?

There is a lot to be learned from this exercise!

Notes

What went right:

_____
_____
_____
_____

What went wrong:

_____
_____
_____
_____

What to do differently next time:

_____
_____
_____
_____

Figure 35-1

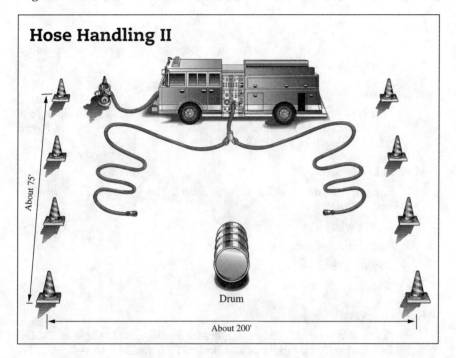

36. Forcible Entry—Through-the-Lock Methods

Objective

To develop proficiency in opening door locks for reasons of safety, damage control, and good public relations.

Setup Time

This drill requires visiting locksmiths, locating locks, and mounting the locks. You should get some help in assembling the props about a month before you schedule this drill.

Materials Required

Fire Engineering's Forcible Entry Video #2, *Through the Lock: Cylinders and Key Tools.*

A door with locks.

An Adams Rite lock.

Shove knives.

K tool.

References

"Lock Picking and Emergency Entry for Firefighters, Police, and EMTs," Bob Karber, High Park Lock and Design, P.O. Box 200516, Arlington, TX 76006, 416/604-8065.

"Forcing the Tubular Dead Bolt," Bill Gustin, *Fire Engineering*, January 1996, page 89.

"Forcible Entry," Tom Brennan and Paul McFadden, *Fire Engineering*, May 1984, page 36; March 1985, page 30; January 1986, page 18.

"Forcible Entry—Through-the-Lock Method," K Tool Manufacturing Company, P.O. Box 58, 14 Oak Spring Road, West Nyack, NY 10994.

Preparation

1. This drill takes some time to prepare. Start by obtaining a copy of the reference book by Bob Karber.
2. Visit your local locksmiths and see if you can obtain some of their samples, which are small sections of doors with locks mounted on them. You can also get an old door and mount several locks of different types along the edges.
3. Get at least one Adams Rite lock mechanism and mount it in a door or in a section of 2 × 6 as shown in Figure 36-1.
4. Locate a door in the fire station where you can demonstrate and have members practice use of the shove knife.
5. Try to schedule this drill just ahead of an auto extrication drill so that you can practice opening the doors of a donated car with a slim jim. Do not practice on either a member's or a departmental vehicle. Some door lock mechanisms have linkages designed to come apart if a slim jim is used (Nissan Sentras in particular will drop the linkage).

Predrill Briefing

"There are any number of occasions when you have to get into a building or apartment, but there is no real evidence to justify doing major damage to effect entry. Lockouts with food on the stove and children left inside are examples of situations requiring entry that is cosmetically correct. You can also create a good public image if you cause no major or expensive damage to the structure. The purpose of this drill is to review some of those methods."

Running the Drill

Part I—Shove Knives

1. Demonstrate the function of the lock bolt. Show how it prevents the bolt from being slipped.
2. Size up the bolt. If there is no dead bolt, the shove knife may be viable.
3. Demonstrate how to use the shove knife. Note that spreading the frame slightly in conjunction with using the shove knife may slip the bolt.

Part II—Videotape

Part III—Practical

1. Remove the cylinder from the Adams Rite lock.
   (a) Show how fine the threads are.
   (b) Show how small the set screw is.
   (c) Emphasize how expensive the door is and the hazards to firefighters.
   (d) Practice with the key tool.
2. Remove the cylinder from the high-security lock.
   (a) Show the size of the set screw—it's harder to twist out.
   (b) Note that, with a double lock, you have to work two slides with the key tool.
3. Rim lock.
   (a) Look at the back of the lock.
   (b) Fit the key tool to trip the bolt.
   (c) Drive off the lock if the night latch is set.
4. Picking locks.
   (a) Picking locks takes a lot of practice.
   (b) The principles of picking locks.
   (c) What tools to use.
5. Opening car door locks.

   Using a donated vehicle, have the members practice unlocking the doors with a slim jim.

Volunteer Training Drills

Notes

What went right:

_____
_____
_____
_____

What went wrong:

_____
_____
_____
_____

What to do differently next time:

_____
_____
_____
_____

Figure 36-1

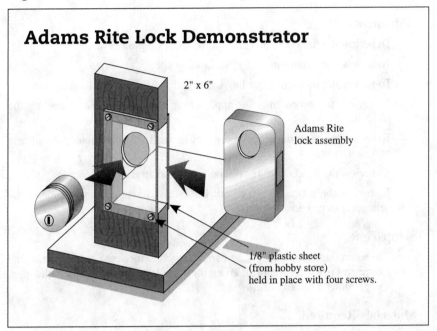

37. Medevac Helicopter Operations

Objectives

To review the protocols and criteria for requesting medevac.

To review the requirements for the landing site.

To review the protocol regarding lights and marking of the landing site.

To review the procedures for approaching the helicopter and loading the patients.

To review the procedures for emergency access to the helicopter crew in case of an accident.

To review the procedures for disconnecting batteries.

To review the procedures for your EMTs and paramedics who ride in the helicopter to provide support.

Setup Time

The setup time involves making contact with the medevac agency, arranging a time and place for the drill, arranging for a landing site for the helicopter, and planning a backup drill.

Materials Required

Any materials requested by the medevac crew and those that are required for your backup drill.

References

The local SOP governing medevac operations.

Preparation

1. Contact your medevac helicopter representative to schedule this drill. If at all possible, arrange for the helicopter to come to your location. Be sure to have a backup drill ready to go in case the helicopter is diverted to an incident.

2. If you intend to use an athletic field, be sure to obtain permission from the proper parties.

Running the Drill

1. If helicopters are going to land at a local field, arrive at the field early. Survey the area for any debris that could be drawn into the engines or thrown about by the rotor wash. Place members around the field to prevent children from entering the landing area. Caution any civilians about the potential eye hazards of dust and debris during takeoffs and landings.

2. Introduce the medevac agency representative and turn the drill over to him.

Debriefing

Base the debriefing on your observations and the input of the medevac personnel.

Notes

What went right:

_____
_____
_____
_____

What went wrong:

_____
_____
_____
_____

What to do differently next time:

_____
_____
_____
_____

38. Burns

Objective

To review the protocol for estimating the degree and coverage of burns.

To review the protocol for referring burn patients.

To review the protocol for treating burns.

Setup Time

30 minutes.

Materials Required

Copies of the attached pretest.

Overhead transparency of patient outline.

Overhead projector and screen.

Pencils for each member.

References

Essentials of Emergency Care, David Limmer, Bob Elling, Michael O'Keefe, and Edward T. Dickinson, M.D., Prentice Hall Inc., 1996.

"Emergency Care and Transportation of the Sick and Injured," American Academy of Orthopaedic Surgeons.

Preparation

Review your local protocols and take the following test yourself. Be sure that you know the answers!

Running the Drill

1. Start the drill by having the members take the pretest.
2. Review the answers to the test.
3. Project the patient outlines on the screen. Review the percentage assigned to each burn on each area of the body.
4. Select members to describe the treatment of a first-, second-, and third-degree burn.
5. Review the protocols for referral to a burn center.

Drill Outlines

Notes

What went right:

_____
_____
_____
_____

What went wrong:

_____
_____
_____
_____

What to do differently next time:

_____
_____
_____
_____

Volunteer Training Drills

Figure 38-1

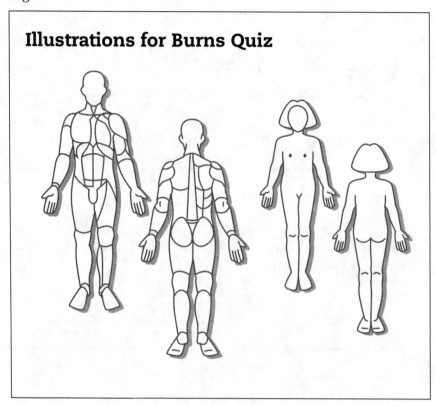

(Figure 38-2)

Burns Quiz

1. The terms first-, second-, and third-degree burns refer to what characteristic of a burn?

   _____

   _____

2. Describe the signs and symptoms of a first-degree burn.

   _____

   _____

3. Describe the signs and symptoms of a second-degree burn.

   _____

   _____

4. Describe the signs and symptoms of a third-degree burn.

   _____

   _____

5. Using the diagrams below, show the percentage of body area associated with each burn.

   _____

   _____

6. List the criteria for referring a burn patient to a burn center.

   _____

   _____

Figure 38-3

Quiz Answers

1. The depth of the burn to the skin.
2. Reddening of the skin with some swelling.
3. Deep pain. The skin will be red with blisters, and the skin may appear to have spots.
4. The pain will range from severe to none, depending on the amount of nerve damage. There will usually be some black charring. Some areas may be dry and white.
5. Use the rule of nines:

	Adult	**Child**
Head	9%	18%
Arm—front	4½%	4½%
Arm—back	4½%	4½%
Trunk—front	18%	18%
Trunk—back	18%	18%
Leg—front	9%	7%
Leg—back	9%	7%

6. Refer to local protocol.

39. Fractures

Objective

To practice the techniques for splinting fractures.

Materials Required

Splinting materials.

References

Essentials of Emergency Care, David Limmer, Bob Elling, Michael O'Keefe, and Edward T. Dickinson, M.D., Prentice-Hall Inc., 1996.

"Emergency Care and Transportation of the Sick and Injured," American Academy of Orthopaedic Surgeons.

Preparation

Arrange a work area and assemble splints and bandages.

Running the Drill

1. Assign the members to work in pairs.
2. The members should:
   (a) Assess the airway, breathing, and circulation of the victim. Be sure that they take a distal pulse of an injured limb.
   (b) Demonstrate the ability to immobilize and treat for (1) fractured forearm, (2) dislocated shoulder, (3) dislocated elbow, (4) fractured pelvis, (5) fractured femur, (6) fractured tibia, and (7) fractured ankle. Be sure that the members assess the distal pulse after the bandage and splints are in place.

Volunteer Training Drills

Notes

What went right:

_____
_____
_____

What went wrong:

_____
_____
_____
_____

What to do differently next time:

_____
_____
_____
_____

40. Soft Tissue and Penetrating Wounds

Objectives
To practice the techniques for bandaging injuries.

To review the protocols for application of the pneumatic antishock garment.

Setup Time
One hour to review the local protocols.

Materials Required
Bandage materials.

References
Essentials of Emergency Care, David Limmer, Bob Elling, Michael O'Keefe, and Edward T. Dickinson, M.D., Prentice-Hall Inc., 1996.

"Emergency Care and Transportation of the Sick and Injured," American Academy of Orthopaedic Surgeons.

Running the Drill
1. Assign the members to work in pairs.
2. The members should:
   (a) Assess the airway, breathing, and circulation of the victim. Be sure that they take a distal pulse of an injured limb.
   (b) Demonstrate the ability to manage and treat for (1) arterial bleeding from the thigh, (2) arterial bleeding from the arm, (3) a gunshot wound to the chest, (4) a steel rod penetrating and extending from the chest, (5) an evisceration of the abdomen, and (6) a pencil protruding from the eye. Be sure that the members assess both entrance and exit wound sites for penetrating injuries.
3. Since penetrating wounds present the potential for large blood loss, review the criteria for application of a pneumatic antishock garment. Be sure to review the signs and symptoms that would indicate the garment should not be applied.

Volunteer Training Drills

Notes

What went right:

_____
_____
_____
_____

What went wrong:

_____
_____
_____
_____

What to do differently next time:

_____
_____
_____
_____

41. Cold Weather Emergencies

Objective
To review the protocols and procedures for treating injuries and illnesses associated with cold weather.

Setup Time
One hour to review the local protocols.

Materials Required
Local protocols.

Locally available transparencies.

Copies of the pretest for each member.

Pencils for each member.

References
Essentials of Emergency Care, David Limmer, Bob Elling, Michael O'Keefe, and Edward T. Dickinson, M.D., Prentice-Hall Inc., 1996.

"Emergency Care and Transportation of the Sick and Injured," American Academy of Orthopaedic Surgeons.

Preparation
Review the pretest. Make sure that you know the answers yourself!

Running the Drill
1. Provide the members with the attached pretest.
2. Following the pretest, review the following subjects: (1) the signs and symptoms of mild hypothermia, (2) the treatment protocols for mild hypothermia, (3) the signs and symptoms of severe hypothermia, (4) the treatment protocols for severe hypothermia, (5) the signs and symptoms of frostbite, and (6) the treatment protocols for frostbite.
3. Review the procedures for responding to an ice rescue. See Drill 25, Drowning Response.

Volunteer Training Drills

Notes

What went right:

_____
_____
_____
_____

What went wrong:

_____
_____
_____
_____

What to do differently next time:

_____
_____
_____
_____

Figure 41-1

Cold Weather Emergencies Quiz

1. Fill in the following table:

Condition	Frostnip	Frostbite	Freezing
Skin surface			
Tissue under the skin			
Skin color			

2. Patients are often unaware that they are suffering from frostbite. **True False**

3. What is the difference between frostnip and frostbite?
   _____
   _____

4. Describe the emergency care for frostnip.
   _____
   _____

5. Describe the difference between superficial and deep frostbite.
   _____
   _____

6. Describe the emergency care for frostbite.
   _____
   _____
   _____

7. Describe the signs and symptoms of hypothermia.
   _____
   _____

Volunteer Training Drills

8. Describe the emergency care for mild hypothermia.
   _____
   _____
   _____

9. Describe the emergency care for severe hypothermia.
   _____
   _____

10. Describe the signs, symptoms, and treatment of trench foot.
    _____
    _____

11. Describe the actions to be taken to protect accident victims from the cold.
    _____
    _____

Figure 41-2

Quiz Answers

1. Fill in the following table:

	Frostnip	Frostbite	Freezing
Skin surface	Soft	Hard	Hard
Tissue under the skin	Soft	Soft	Hard
Skin color	Red, then white	White, waxy	Blotch white to yellow gray, then blue gray

2. True

3. Frostnip is not serious and the tissue damage is minor. Frostbite involves the skin and subcutaneous layers.

4. Allow the patient to warm the part by applying warmth from his own hands; otherwise, blow warm air on the site.

5. With superficial frostbite, the skin appears white and waxy. Only the surface feels frozen and the skin below the surface still feels soft. With deep frostbite, the tissues feel frozen and are no longer resilient. The skin color is mottled or blotchy and turns from white to yellow gray, then blue gray.

6. Rewarm the body part by submerging it in 100- to 105-degree water without letting the body part contact the sides or bottom of the container. After warming, keep the site warm according to local protocols.

7. Shivering, feeling of numbness; drowsiness and decreased level of consciousness; slow breathing and pulse rate.

8. Keep the patient dry. Slowly warm the body core before warming the extremities. Keep the patient at rest. Treat for shock and apply oxygen. If alert, give warm fluids.

9. Handle with extreme care to avoid ventricular fibrillation. Provide oxygen, warmed if possible. Wrap the patient in blankets and transport immediately.

10. The limb is swollen, with a waxy and mottled appearance, and it is cold to the touch. Remove wet shoes and socks. Gently rewarm the extremity and wrap it with a loose, sterile dressing. Transport.

11. Provide protection from wind and cold using blankets, salvage covers, or plastic trash bags. Cover the head to prevent heat loss. Apply oxygen from a warm oxygen bottle.

42. Hot Weather Emergencies

Objective
To review the protocols and procedures for treating injuries and illnesses associated with hot weather.

Setup Time
One hour to review the local protocols.

Materials Required
Local protocols.

Locally available transparencies.

Copies of the pretest for each member.

Pencils for each member.

References
Essentials of Emergency Care, David Limmer, Bob Elling, Michael O'Keefe, and Edward T. Dickinson, M.D., Prentice-Hall Inc., 1996.

"Emergency Care and Transportation of the Sick and Injured," American Academy of Orthopaedic Surgeons.

Preparation
Review the pretest. Make sure that you know the answers yourself!

Running the Drill
1. Provide the members with the pretest.
2. Following the pretest, review the following subjects: (1) the signs and symptoms of heat cramps, (2) the treatment protocols for heat cramps, (3) the signs and symptoms of heat exhaustion, (4) the treatment protocols for heat exhaustion, (5) the signs and symptoms of heat stroke, and (6) the treatment protocols for heat stroke.
3. Review the procedures for monitoring firefighters during suppression operations or extended operations in hot weather. Review rehab procedures. Emphasize the need for officers to be proactive in evaluating members and not just ask them if they are feeling okay.

Notes

What went right:

_____
_____
_____
_____

What went wrong:

_____
_____
_____
_____

What to do differently next time:

_____
_____
_____
_____

Figure 42-1

Hot Weather Emergencies Quiz

1. Fill in the following table:

Condition	Heat cramps	Heat exhaustion	Heat stroke
Muscle cramps (yes/no)			
Breathing characteristics			
Pulse rate			
Weakness (yes/no)			
Skin moisture			
Muscle cramps			
Perspiration			
Loss of consciousness			

2. Describe the emergency care procedures for heat cramps.
3. Describe the emergency care procedures for heat exhaustion.
4. Describe the emergency care procedures for heat stroke.

Figure 42-2

QUIZ ANSWERS

1. Table Answers

Condition	Heat cramps	Heat exhaustion	Heat stroke
Muscle cramps (yes/no)	No	No	Yes
Breathing characteristics	Varies	Shallow, rapid	Deep, then shallow
Pulse rate	Varies	Weak	Full and rapid
Weakness (yes/no)	Yes	Yes	Yes
Skin moisture	Normal, moist	Cold, clammy	Hot and dry
Muscle cramps			
Perspiration	Heavy	Heavy	Little to none
Loss of consciousness	Not often	Occasionally	Likely

2. Move the patient to a cool place. Gently massage the cramped muscles. Give diluted electrolyte fluids.

3. Move the patient to a cool place. Remove the clothing to allow fanning of the skin. Be careful not to cause chills. Give diluted electrolyte fluids. Treat for shock.

4. Immediately cool the patient by any means possible. Wrap in wet sheets and apply continuous cooling. Use cold packs at the armpits, at the wrists, and behind the knees. Provide emergency transport.